The Deltoid Pumpkin Seed

BY JOHN McPHEE

The Founding Fish
Annals of the Former World
Irons in the Fire
The Ransom of Russian Art
Assembling California
Looking for a Ship
The Control of Nature
Rising from the Plains
Heirs of General Practice
Table of Contents
La Place de la Concorde Suisse
In Suspect Terrain
Basin and Range
Giving Good Weight
Coming into the Country
The Survival of the Bark Canoe
Pieces of the Frame
The Curve of Binding Energy
The Deltoid Pumpkin Seed
Encounters with the Archdruid
The Crofter and the Laird
Levels of the Game
A Roomful of Hovings
The Pine Barrens
Oranges
The Headmaster
A Sense of Where You Are

The John McPhee Reader
The Second John McPhee Reader

THE DELTOID
PUMPKIN SEED

John McPhee

Farrar, Straus and Giroux
New York

Farrar, Straus and Giroux
18 West 18th Street, New York 10011

The text of this book originally appeared in The New Yorker,
and was developed with the editorial counsel of William Shawn
and Robert Bingham.

Library of Congress catalog card number: 72-84783
ISBN: 0-374-51635-9
EAN: 978-0-374-51635-2

Designed by Paula Weiner

www.fsgbooks.com

30 29 28 27 26 25 24 23

FOR JENNY

The Deltoid Pumpkin Seed

I N G R E A T S E C R E C Y, on a private airstrip about
fifty miles southwest of New York, Aereon 7 got ready
to fly. Conditions were good. It was a clear August evening in 1970, humid, but not remarkably so for that time
of year near the coastal plain. In dark stands of ash and oak
by the field, leaves were not moving. A wind sock, almost
half a mile away, hung still; and far beyond that, on the
horizon of this flat landscape, stood barracks of the state
police, where no activity was discernible, and where the
flags of New Jersey and the United States hung without
motion from tall poles.

The 7, as the aircraft was called, was bright orange. It
had no wings. It had a deep belly and a broad, arching
back. Seen from above, it was a delta. From the side, it
looked like a fat and tremendous pumpkin seed. Stabilizing
fins, vertical and anhedral, framed the trailing edge. The
aircraft's shape had been figured out by a computer in

Valley Forge, Pennsylvania, and the computer had sought a shrewd and practical compromise between an airfoil and a sphere. The 7 stood on tricycle landing gear. Above and just behind its nose was a plastic canopy. Mounted above the aircraft's trailing edge was an engine fitted with a pusher propeller. Aereon 7 had cost about five thousand dollars. To reach this moment, though, well over a million dollars had been spent in the past eleven years. The money had been drawn from various individuals, many of them in New Jersey, in amounts ranging from five hundred dollars to three hundred thousand dollars. Nothing had been contributed by the government or by any major aircraft corporation. By training, the several men attending the aircraft were all engineers (aeronautical and mechanical) except one, and he was a Master of Theology. His name was William Miller, and he was of middle height and middle weight and had a look in his clear blue eyes that all at once seemed to be prayerful, patient, lonely, trusting, and nervous. Except in technical matters, the others deferred to him. He had brought them together, and they were in his employ.

The engine started and began to thrum. The 7 vibrated, suggesting an airliner getting ready to leave its bay. For two or three minutes, it stayed where it was, near the edge of the field, then it turned slowly and began to taxi, under remote control. It seemed to grow as it rolled. In that milieu—the deserted airstrip, the flat expanse, the complete absence of all other aircraft anywhere in sight—perspective went where it pleased. The 7 was exactly seven feet long, but, taxiing out there by itself, with no human being anywhere near it, it took on added proportion until it seemed huge—as huge as its progeny were intended to be. It waddled toward the head of the runway. Its flaps moved up and down in final checkout, its rudders back and forth. The 7 had flown a number of times before this outing, but not smoothly. It had a tendency to jackrabbit, to fly in short oscillations that frequently intersected the ground, to bounce, to bounce again, and bounce again, and flop, and skid on its nose. It had never been in a state of steady flight. It had not yet attained an altitude above fifty inches. Now, turning on the broad white stripes at the head of the runway, it sat for a while, pointing toward the far end, waiting for clearance. Then the sound of the engine rose, and the 7 began to move. Smoothly, steadily, its bulbous frame collected speed. Its engine roar cut up the summer air. After three hundred feet, the aircraft rotated around its center of gravity, gently lifting its nose wheel off the ground. Then, after five hundred feet, it firmly took to the air. It climbed out nicely to an altitude of a hundred inches, threw one blade of its propeller, sank rapidly, landed heavily, and scraped its nose. The propeller was replaced and the 7 made a few fast runs without lift-off, but there was no more flying that day. Data had been collected, despite the accident, and the data were to be analyzed before further testing. Miller said, "We want to be able to use the 7 again and not lose it in some spectacular display of virtuosity. We still have a lot to learn from it." The aircraft taxied in, wheeled around, and stopped. Its engine shut down. Debriefing, the engineers spilled their thoughts—totally technical—into a cassette recorder.

In a hangar at an even remoter airfield, some thirty-five miles farther down into southern New Jersey, was Aereon

7's fraternal twin, Aereon 26. With Sheetrock and two-by-fours, the 26 had been hidden away in a huge box that filled a large part of the hangar—this to thwart not only a paid spy but also any innocent but curious observer who, for example, might say to friends, "My God, you'll never believe what I saw today at Red Lion Airport! They've got a plane there, a big orange thing, with no wings." Stories like that tend to expand and to travel, and that is what Miller deeply feared. Aereon 26 had been built in a small shop near Lakehurst, New Jersey, by a retired airship rigger named Everett Linkenhoker. He built the aircraft in two parts, and eventually he secured them, one at a time, to the roof of his station wagon and drove them in the dead of night to Red Lion, where he put the two halves together and hid the 26 in the Sheetrock box. The advantage of Red Lion was its obscurity. Over a period of many months, various taxi tests were conducted on the runway there, always at odd hours—just before dusk, or in the early morning. A car towed the 26 around while friction drag was measured with a hook-and-eye scale of the type that is used to weigh swordfish.

There had been other Aereons. In existence still, in a basement in Trenton, was an Aereon twenty inches long. There had been a series of four-foot Aereons. Now there were the 7 and the 26; and as yet unbuilt, but much alive in Miller's imagination, were a fifty-two-foot Aereon, a two-hundred-foot Aereon, a three-hundred-and-forty-foot Aereon, and an Aereon whose length over all might approach a thousand feet. But the 26 was the key vehicle. It would nourish or finish the Aereon project. It would embarrass or vindicate the computer in Valley Forge. It would fly, or try to, with a pilot under its canopy. His name was John Olcott. Because Aereon 7 was now nearing the end of its test series, the time had at last come to take the 26 out of concealment at Red Lion, where the runway was not adequate for flight tests, and, risking exposure, move the vehicle to a major airfield.

Even farther down into New Jersey, off the southern edge of the Pine Barrens, is an airfield more or less the size of Kennedy International. Planes of all United States airlines go in and out of there, but without passengers. The place is known as NAFEC—an acronym that long ago swallowed its own meaning—and NAFEC is where the Federal Aviation Administration tests and evaluates aircraft of all sizes and types. August 13, 1970, at Red Lion, Aereon 26 was chained to the bed of a tractor-trailer ordinarily used to transport bulldozers. The aircraft's markings were covered by sheets of newspaper stuck on with masking tape. The engine and propeller were sheathed with canvas, and the cockpit's plastic bubble was masked out, too. As a wide load, the rig was not permitted to move on New Jersey roads in darkness, so it left Red Lion Airport the following morning at dawn, scraping bushes and limbs on narrow blacktop roads, and eventually moving south on Route 206 to Hammonton Circle and east on the White Horse Pike to NAFEC—forty miles, at speeds ranging from ten to twenty miles per hour. Miller shepherded the 26 in his Mercedes-Benz. As the journey progressed, he befrazzled himself with worry that sinister eyes would fall upon his creation.

The main building at NAFEC is immense, with sliding steel doors seventy feet high closing the ends of a room in which several big jets and any number of small planes can be housed at the same time. The floor personnel there have worked on just about everything, military or civilian, that flies in American sky. It is hard to imagine any sort of aircraft that would draw from them more than a glance. Around nine o'clock in the morning, though, on that August day, when Aereon 26 came in and was lifted off the flatbed trailer, men from every part of the hangar left the planes they were working on and collected in a wide circle around it. They looked at it in silence. In time, they would be making cynical remarks about the wingless orange vehicle and its trials, but now they just stood there and took it in. The 26 was placed at one side of the hangar. Two fifty-pound weights were hung from its nose, because its balance was so critical that, without a pilot inside it, the 26 would otherwise have canted backward. Stanchions were put around it. It was roped off like an exhibit in a museum. As the NAFEC men watched, the Trenton *Times* was peeled away from the orange fuselage. A big sheet came off the side revealing black registration numbers, "N2627." More newspaper came off the underbelly, introducing to the NAFEC people the name "AEREON 26." And still more news came off the vertical fins, where black lettering said "AEREON AEROBODY 001."

Somewhere in the administrative modules of the big building, a telephone rang, and a motorist who had been out on the White Horse Pike asked what was that big orange saucer that had gone into NAFEC by truck. The call was referred to McGuire Air Force Base, which is on the far side of the Pine Barrens, fifty miles away—609-724-

2100. The switchboard at McGuire is one of the darker jungles in the topography of communication. Call after call goes in there and is never heard from again. The query about the Aereon died there.

A BOUT THE MAN who would get into the 26 and try to fly it there was something studiously, insistently Hessian. He was, in a sense, a hired gun. He had no emotional attachment to the Aereon project. He had no financial interest in the Aereon Corporation. His technical curiosity had been aroused just enough to cause him to accept the job. He was working as an independent contractor. He minimized the risk, because, for one thing, he believed in the computer. He was no stunt man. He was an aeronautical engineer, a member of a firm of consulting engineers, and his central interests were in things like equations of motion and what they might tell him about the performances of vehicles in the air. He said, "We're exploring relatively unknown areas, it's true. The vehicle is not an airship, and it is not an airplane. It is a hybrid of the two, trying to combine the benefits of heavier-than-air aerodynamics and lighter-than-air aerostatics—a vehicle that would have the capability of carrying large volumes of cargo without the operating problems normally associated with airships. A hybrid vehicle like this may have some unusual characteristics, but a pilot flies in an adaptive manner. He observes. He adapts. He achieves what he desires to achieve." Olcott hardly appeared to be the sort of creature who dances around on the lips of danger seeking the pleasures of not being swallowed. About six feet tall, trim and lithe, handsome and blue-eyed, with his hair cut short and precisely parted and combed, he looked like the president of anybody's student council some years ago. He always wore a tie and a button-down shirt and generally a plain suit, but sometimes, in this summer weather, a madras jacket. He spoke quietly, in an earnest but—at least within sight of an airfield—never an agitated voice. He seemed, in the main, to be so completely two-dimensional that he might have been cut out of a newspaper, except when he lost contact with the others before a test or a debriefing—as he did from time to time—and went off into his own mind somewhere, while his eyes shot the horizon blankly or stared into a floor.

Olcott was thirty-five, and he had been flying since 1951, when he was fourteen and made the discovery that a fourteen-year-old could be licensed to fly gliders in New Jersey. His father was an electrical engineer who feared flying so much that he took trains on business trips. Nonetheless, he drove his son sometimes twice a weekend the fifty-mile round trip from their home, in Short Hills, to Somerset Airport. Gliding gives a pilot the fundamentals in a way more basic than any other. This is probably why the pilots of the Luftwaffe, in their time, were the world's best: they had learned their flying in gliders, because the Versailles Treaty denied them the experience of powered flight. Olcott, later on, would assign particular value to the time he had spent hunting the air for thermals. "It's good discipline," he would say. "You have to be right the first time. There is no margin for error, because you don't have an engine. So you develop a more keen awareness of the vehicle in its environment." He flew gliders over Switzerland at the age of seventeen, and in the same year got his license as a pilot of powered aircraft. He had his commercial license at eighteen and his instructor's license at nineteen. He went to Princeton, earning his undergraduate and graduate degrees in aeronautical engineering, and he flew charter flights whenever he could to supplement his income and to build up his log of flying hours. When he agreed to be the test pilot of the Aereon 26, he had well over five thousand hours in the air—a lot for a desk-job engineer. Unusual test programs were not unusual to him. He had once spent almost two years at the Indian Institute of Technology, in Uttar Pradesh, where he tested for the Indian government prototype aircraft of Indian design. Why Miller had hired him to test the 26 was in part a result of propinquity. Olcott's daylight work was at Aeronautical Research Associates of Princeton, Inc., in Princeton Junction, and a couple of rooms above a bank on Nassau Street in Princeton happened to be the home and only offices of the Aereon Corporation.

In his madras jacket, a blue button-down shirt, and a knit tie, Olcott appeared at the Princeton University airstrip on another calm, hot evening toward the end of August, for what would prove to be the penultimate test of the

7. He did not feel altogether prepared to get into the big Aereon and take stick in hand, down at NAFEC, until he had further studied the behavior of the smaller Aereon in flight under remote control. This time, Olcott sat in the right front seat of a Buick convertible, its top down, visibility optimum. The plan was that the car would race on the runway in tight formation with Aereon 7, the better to afford close observation. Miller, with a Super 8 motion-picture camera in his hand, sat directly behind Olcott. The car was otherwise stuffed with engineers, one of whom, John Kukon, held a Logictrol transmitter that would move the 7's control surfaces.

"O.K., let's get this airplane off the ground," someone said.

Miller—a gentle bit of a nag—reminded the group as a whole that they were to refer to the aircraft as an "aerobody." He said, "I'm very anxious that we preserve this semantic thing. This is not an airplane. We consider 'aerobody' to be a generic description we have coined. That's well understood, isn't it?"

"Roger. The thing doesn't know whether it's an airship or an airplane."

Miller seemed content with the definition. On both sides, there was a tenuous balance, a conscious application of patience, a sense of tolerable difference, in the relationship between Miller and these engineers. While Miller knew more than most laymen about structural analysis and flight mechanics, he did not know much more, and when he was among his assembled consultants he was not fully in the conversation. They politely ignored him much of the time, and they seldom showed annoyance when he fussed and worried. They went along with him when he prayed. They were interested in the project, and they knew that beyond question there would be no project at all, by now, were it not for him. The company had been on a low threshold for so long that its mere continuance was marvellous. It had been humbled with accident and with failure. It had been strictured by the Securities and Exchange Commission. When Miller had emerged as president, the board of directors had grown even more nervous than it had been before, but Miller was the only chrysalis willing to come out for the job. Aereon now was costing fifteen thousand dollars a month. Miller was somehow engendering the money. He worked alone—no secretary, a part-time accountant. The engineers knew, but not from Miller, that Miller had put something like three hundred thousand dollars of his own money into Aereon, almost totally evaporating his inheritance, his portfolio, and even his Navy flight pay, every cent of which he had saved.

Before Miller went to college, as a late beginner, and on to divinity school, he had flown Skyraiders off the carrier Kearsarge. He also flew Helldivers, Corsairs, and Grumman Avengers. In his boyhood, there had been two main directions in which he had thought he might go. He wanted to be a pilot, and he wanted to be a missionary. He was born in 1926 in Meshed, a sacred Muslim city in northeastern Iran, and he used to hike in the Elburz Mountains and watch birds drifting in the thermals. He collected feathers and stuck them in his jacket. Airplanes were scarce in Persia. When Miller saw one, he felt an impulse to run to the mountains and get up on top of them and reach up and touch. His father was a missionary. His mother's family was focussed on an estate called Wyck, in Philadelphia, which had been in the family for three hundred years. Miller lived there for a time also, and went on to Choate and to Princeton. Before being drawn into complete absorption in Aereon, he had done church work of various kinds, but he had never sought a church of his own, or been ordained. He now lived alone in a garden apartment behind a commercial complex on U.S. 1 northeast of Trenton. For three years, he had been working six days a week, twelve and fifteen hours a day, to preserve Aereon's existence, and beyond this the only—even mild—suggestion of self-indulgence among his habits, choices, or customs was the Mercedes-Benz that was parked beside the airfield. On the Mercedes' back seat was a blue Frisbee, its perimeter decorated with pictures of Jupiter, Saturn, Uranus, Neptune, Pluto, Mercury, Venus, Earth, and Mars. At airfields, Miller threw the Frisbee when conditions were not perfect for his Aereons.

After four high-speed taxi runs, the 7 was ready to fly. The slanting rays of the late sun got inside the aerobody and lit up its structure from within—luminescent slides of orange on a trellis of bowing ribs. In the convertible, in the driver's seat, was William Putman, an aerodynamicist on the faculty of Princeton University, whose particular skills had to do with low-speed aircraft. He was wearing a sports shirt that appeared to have been cut from a gold-rouge-and-green awning. "In gear," he said. He was driving because it was his car. Like many aeronautical engineers, he was not a pilot. The only vehicle in which he had ever addressed the head of a runway was his Buick.

There was a brief spray of last-minute dialogue.

"Should we try a few runs to practice coördinating pitch control with power application?"

"No. We're too close to the ground to try that. We're likely to pop off into some bad attitudes."

"This one is rotation with lift-off and sit-down—no lateral inputs."

"Let's not accelerate too fast."

"The gentler the better."

"No matter how it goes, we must not get false confidence from the 7. Remember, there's going to be a human being in the 26."

Olcott said, "Let's try to get a feeling for what coördination of throttle and stick is required to hold a constant pitch attitude."

Then, side by side, the automobile and the aerobody began to move. The two engines crescendoed. "Twenty-five!" Putman called out. "Thirty! Thirty-three!" The 7's nose came up and held steady in its angle as the aircraft and the car gained parallel speed.

"I'm a little chicken about getting too much closer to that thing," Putman said.

Olcott, beside him, said, "I would suggest you get a little farther away from it."

"Thirty-four!"

"How's that engine?"

"It's good."

"Thirty-five!"

The 7 went into the air at thirty-five miles per hour and climbed until it was eight inches off the ground. Then it began to skip like a stone—four times. Then, still gaining speed, it went up a little higher, and into further oscillation.

"Forty!"

The 7 bounced hard four more times, suddenly veered, closed in toward the Buick, tried to cross in front of it, and —in a nauseating screech of brakes—was crushed. Everybody jumped out of the car. In an atmosphere of absolute calm, they began at once to debrief. Miller himself showed no emotion. He circled the wreckage and recorded it with his Super 8. The right-front tire had crunched far into the left side of the 7, and the notch the tire had made was edged with shattered superstructure and torn orange silk. The debriefing established that the 7 had not turned in the air. The accident had not resulted from an aberration in flight. The 7 had, in fact, been glued down to the runway after the series of bounces, and had just begun its roll-out—its deceleration toward a stop—when the nose wheel had gone out of control. Olcott showed no alarm, no dismay, and no particular interest in the accident or its cause. Instead, he was absorbed with things he had noticed while the aerobody was scuffing along the runway in its awkward bursts of flight. "It occurred to me that the type of oscillation it got into was the type of oscillation that could not be controlled before. But it was under control this time," he said.

"That's right."

"We don't really know the air speed very well. We may be five knots slower than we anticipate, and five knots is very important. I think we've got to be right on the precise air speed or we're going to have response problems.

I believe that the computer simulation and what we're seeing here with the model points up the necessity for very precise air-speed control."

"Are we going to repair this thing, and, if so, how will that affect the test schedule of the 26?" Putman said.

"It will delay it," Miller said reflectively. Keeping the 26 at NAFEC was costing about fifty dollars a day.

"From a technical point of view, we haven't had a successful outing with the 7," Olcott said. "We can argue amongst ourselves that the problems we've been experiencing are unique to the 7, and won't be transferred to the 26. But we can't say it with a high degree of certainty. The 7 may be trying to tell us something."

John Kukon lifted a flap of the broken silk skin of the 7 and looked at the destruction inside. He said, "I think we ought to try to get this thing at least twenty-five feet up in the air before the 26 goes. I really do."

There was a period of silence, during which everyone looked contemplatively at Kukon.

"How long will it take, John?"

Kukon had made the 7 in his basement. He was a master builder of aircraft models—with a virtuosity few other people had ever approached.

"Flat out—working on it myself—it will take a week," he said. "There's a fair amount of work involved. Even the main gear struts are broken up."

"What will it cost, John?"

"Three hundred to five hundred dollars."

Kukon spoke in a high nasal intonation, and he spoke rapidly—like a recorded voice going too fast but not unintelligibly so. Everything about him seemed quick. His eyes, dark brown, moved quickly. His gestures were staccato. His face, under wet-combed dark hair, was flushed. He was sick with a virus and should have been in bed.

"Would it go faster if you had help, John?"

"From somebody who knows how to build models, yes. From somebody I have to teach the whole story, there's no hope."

A name was mentioned. "How about him, John? What is your estimate of his ability in building models?"

"If he has a kit, he can glue it together."

A FTER AEREON HAD BEEN a going corporation
for more than eight years, it had not yet flown anything higher than the bounce that might happen when a
model hit a stone. This was around the end of 1967, a particularly depressed area in the company's history. Previ-
ously, in another configuration, there had been an eighty-
foot Aereon that had never left the ground and had rolled
over in a gust of wind, more or less destroying itself. Moving on (more cautiously) into the deltoid form, the company built a four-foot plastic Aereon equipped with a
noisy, powerful little gasoline engine. This version—Aereon
4—displayed absolutely no inclination to fly. It was tested
at Lakehurst Naval Air Station, within a short distance of
the historic swatch of ground where the German rigid airship Hindenburg had burned thirty years before. The
model hobbled all over Lakehurst, and occasionally raced
at high speed, but it never so much as tilted its nose into

the air. Small crowds of investors, directors, and other on-lookers were sometimes present. One observer whispered to Miller, "If I were a stockholder in your company, I'd worry about the engineer." Less than a year earlier, Miller had become Aereon's president, and now as his fortunes melted his spirits subsided, too. He was a study in gloom, in Princeton one day, when a friend of his whose work was also in aeronautics happened to see him at a simulated eighteenth-century tavern called the King's Court, where Miller was having lunch. The friend asked what was the matter. Miller confessed his troubles. "The models go rocketing up and down the field, but they won't even budge off the ground," he said. "They're too heavy, plain and simple, and there are problems with the radio control."

"Call John Kukon," the other man said. "K-u-k-o-n. He builds models for the university. There's no one better. Maybe he'll help you."

Miller called Kukon and introduced himself. Aereon's office was then in a hangar at Mercer County Airport, about ten miles west of Princeton. Kukon's house was not far from there. Sure, he would be happy to come have a talk. Why not after work tonight?

When Kukon kept the appointment, Miller told him of the company's difficulties and showed him sketches of the deltoid pumpkin seed. There were no detailed plans and never had been. Kukon examined the sketches and offered a suggestion. Before continuing the four-foot model series, Aereon might do well to start over again—with a model, say, around two feet long. It would be simpler —it could be rubber-powered—and it would probably serve just as well to indicate in a rudimentary way the performance characteristics of an aircraft shaped like the one in the sketch.

Miller considered the idea, and then asked, "Will you build it?"

"Sure, why not have a try?"

"How long would that take?"

Kukon looked around the hangar, and paused to think, while Miller reflected on the sorry history of his company: formed in the nineteen-fifties, now getting on toward the end of the sixties, its closets cluttered with former presidents, with records of a million spent dollars, and with broken aircraft, in various sizes, that had proved to be penguins all. Now here was still another beginning, almost from scratch—new departure, new delay. What difference, though, could another couple of months make after eight and a half years? He asked again, "How long will you need?"

Kukon said, "Will it be all right if I bring it here the day after tomorrow?"

Kukon was twenty-nine years old. He had started building gasoline-powered flying model airplanes when he was seven. Year after year, he built miscellaneously—indoor models, outdoor models, gliders, stunt models, and combat models that fought in the air. He was fifteen when he decided to specialize. He had joined the Academy of Model Aeronautics—the organization that regulates and administers flying-model competitions of regional and national scope—and his choice was whether to fly figure eights in front of judges or to fly for pure speed. "A stopwatch doesn't lie," he told himself. "It doesn't have any personality." So he elected to concentrate in the field of control-line speed. When he tried a competition for the first time, he placed third among thirty contestants, and he felt drawn to what he was doing as never before. He took a paper route to help pay for materials. Afternoons, almost without exception, he came home from high school and immediately went to the basement to build and build and build—Class A's, Class B's, Class C's, proto-speeds, jets. He lived in Fords, New Jersey, outside Perth Amboy, where control-line exhibitions regularly took place in Waters Stadium on the Fourth of July. Kukon would be in there every year, exhibiting his Ringmasters, his scale-model Cessnas, his little P-51s—in dust and smoke and the rampant noise of V-1 scale-model pulse-jet engines. You could hear them seven miles away. He went to a contest somewhere every Sunday—to Wilmington and Baltimore, to airports and fairgrounds, to naval bases, to the Grumman plant at Bethpage, Long Island. The planes were hand-launched. Kukon's father—a cook at a home for disabled veterans in Menlo Park—was his crewman. Once the plane was in the air, Kukon, holding the control line, would jam his wrist into a yoke (a thing that looked like an oarlock) that was set into the top of a short steel pole. The plane would race around in circles on the end of the line while Kukon, rapidly circling the pole, determined the plane's altitude by manipulating a monoline control system with his left hand. Classes were a matter of engine size, and the bigger the engine the longer the control line—forty-two feet, fifty-two and a half feet, sixty feet, seventy feet. The official stopwatch started after three laps. The timed distance was always a half mile. The planes were required to fly at an altitude of fifteen feet or less. The place to

be, though, was close to the ground, because of a phenomenon known in aeronautics as ground effect. Ground effect, or
the ground cushion, as it is sometimes called, is not fully
understood but is somehow related to wingspan. An airplane in flight—any plane, from a small model to a 747—
is in the ground cushion when its altitude measures less
than the spread of its wings. Sitting on the cushion, the
plane gets added lift without the penalty of drag. So Kukon—risking complete destruction of his models, which
flew well over a hundred miles per hour—learned to fly
them within two inches of the ground. He won three hundred and fifty trophies. He once set two national records
in a single week—one in Maryland and the other at the
New York *Mirror* Model Flying Fair at Floyd Bennett
Field. Two thousand contestants were there, most of them
adults. Kukon was nineteen. Flying started at dawn, and
all day long Kukon won prizes. His Class C plane flew nine
miles per hour faster than the national record. Kukon and
his father went home in their twenty-year-old Chevrolet
with five trophies, five radios, four wristwatches, a set of
tools, two cases of Coca-Cola, a box of silverware, three
wallets, and a cookstove. An accident that happened at
Johnsville Naval Air Development Center, near Philadelphia, caused Kukon to decide to give up control-line speed
flying forever. Sitting on a box with a stopwatch in his
hand, he was monitoring a run by one of his regular
opponents, a doctor from the medical faculty of Temple
University. The engine, prop, landing gear, and other attached parts of the doctor's plane—a two-pound package
in all—broke away from the fuselage and projectiled toward Kukon's head. Just then a young boy stepped in

front of Kukon. At eighteen thousand revolutions per minute, the wild engine went into the boy's kidney, nearly killing him. Kukon cancelled all his competitive plans. He had little time, anyway, for much but work and study. After high school, he had enrolled at the Academy of Aeronautics at LaGuardia Airport, where he got his Airframe and Powerplant credentials, the badge of the licensed mechanic. He had been about to go to work at Newark Airport as a mechanic for American Airlines when a friend told him about a job at Princeton that seemed unbelievable. The university actually paid people to build models. They had a unique test facility called the Long Track, a narrow building seven hundred and sixty feet long, where they worked with designs for low-speed aircraft. Soon Kukon was building models at Princeton which cost contracting companies as much as eighty-five thousand dollars; for example, a four-engine vertical-lift model whose wing could rotate through a ninety-degree arc so that the engines would point forward or up, as the pilot chose. Doing consulting work, he built a model tube train. The tube was two thousand feet long. The train was driven by counter-rotating propellers. It shot through the tube at two hundred miles per hour. For all the diversions of his work, though, Kukon's thoughts, in the months that followed the accident at Johnsville, were drawn repeatedly to one trophy, its image sharp in his mind. It was a staggeringly big trophy, an elaborate gold sculpture in late aerobaroque. It was almost as tall as Kukon—a giant gold cup, fourteen inches from rim to rim, and it had a gold lid, on top of which was a gold airplane. It was given for the over-all best performance at a model contest held annually at Westchester County Airport, in White Plains. The trophy belonged to the Westchester Exchange Club, and not to the winner. There was a provision, however, that should anyone ever win it three years in a row that person could keep it permanently. There had never been a three-time winner. Kukon had won the trophy the year before. He had also won it the year before that. This fact tormented his decision to retire. He kept thinking about that stupendous cup with its gold filigree and its gold airplane. In the end, inevitably, he decided to go back into control-line speed for one more day. When the day came, he moved around the airfield from flying circle to flying circle, from class to class, and put his various planes through their lariat flights, always on the edge of record speeds. There were so many contestants that Kukon had time only for one flight in each class he entered, although the rules technically allowed him three. By late afternoon, though, he was beginning to relax into the expectation that the big trophy was on its way to Fords, for he had flown five events and his times were the best in each. Nothing less would do, because in the contest as a whole there were many events—speed events, free-flight events— and even a single second place could eliminate a contender for the high-point trophy. By the rules, all flying would cease at 6 P.M. At five-fifteen, someone beat Kukon's Class A time by one-tenth of a second. With luck, he could try once more. He put his number in at the judges' table and began the wait for a final turn in the Class A circle. Several others were ahead of him. It seemed likely that the contest would close before he could get into the circle. Half an hour went by, while Kukon chewed the ends of his fingers. The nails had long since been bitten back so far he could not reach them. At ten minutes to six, his number came up. Kukon opened his fuel box. He made his own fuels. Fuel bought in a store was castor oil and alcohol, and Kukon was far past that. In his fuel, only five per cent was alcohol. The rest was high-energy material in various blends—propylene oxide, nitromethane. In the fuel box were four bottles, four blends. Each bottle was wrapped in carpeting. Blend 4 had never been used. Kukon had never actually expected to use it. He had conceived of it as a fuel for a situation of extreme and unusual emergency. Its characteristics were that it would almost certainly destroy the engine that burned it, but meanwhile the engine would develop enough thrust to drive a sparrow to the moon. Kukon entered the circle with his Class A plane and poured Blend 4 into the engine. By the rules of the competition, he had three minutes to get the plane into the air. The engine seemed not so much to start as to explode. Its force immediately broke the propeller. Kukon ran to his equipment box for another propeller, and, with his hands shaking, tried to get it mounted before his time ran out. When he finished the job, he had twenty seconds left. He started the engine again. This time the propeller did not break. The airplane, screaming, bolted into the air. Kukon could barely hold on to the control line. He could not pirouette fast enough to keep pace with the plane. It was all he could do to get his wrist into the yoke. Class A planes are small, as control-line speed models go, and no one in the world had ever flown one a hundred and fifty miles an hour. Kukon's model got right down onto the deck, deep in the ground cushion. An audio tachometer covered the run and showed that the engine was doing thirty thousand revolutions per minute. The plane ate up its half mile at an average speed of 150.013 miles per hour. Then it flew on and on and on. It flew six miles. There is no way to shut off one of those engines in the air. When the plane landed, the motor was extraordinarily hot but undamaged.

Forty-eight hours after his first meeting with Miller, Kukon returned to Mercer County Airport with a twenty-inch Aereon in a cardboard box. Using one-thirty-second-inch balsa sheet—following the sketches, working essentially from scratch, from blueprints that developed in his imagination—he had made a double root rib as a keel, two tip ribs, and two more ribs to complete the structure. He had covered it with tissue paper that had come out of a shoebox. To the trailing edge of the delta he had glued a pair of vertical balsa fins. Unlike its successors—the 7 and the 26—this smallest of Aereons had no anhedrals, no horizontal tail fins. Kukon said, "You don't need them indoors—no gusts." Its motor was a single twenty-inch loop of rubber band. For landing, it had wire skids. And for sheer jazz he had painted a red streak down the axis of the delta, top and bottom. Miller and others gathered around him. Kukon held the wingless aerobody in his hand, gave the rubber two hundred turns, and arrested the pusher prop in his fingers. He looked around the big hangar. The place was full of posts and girders and airplanes, including a two-engine Grumman Gulfstream. Kukon himself had never flown anything larger than a model, and he apparently never would. The few times he had ridden in small aircraft, he had turned gray and become sick. He had once made plans to take flying lessons but changed his mind after a ride in a small plane made him sick for two days. He had never flown in a commercial airliner. He had, however, become an aeronautical engineer. At night, after work at Princeton, he had commuted for several years to the Polytechnic Institute of Brooklyn, where he earned his degree. Now, in the Mercer County hangar, he studied the situation some more, and then he breathed on the elevons and the fins of the Aereon to warp them just so, to work the wood, to set a turning radius for the flight. He let the aerobody go. Its propeller whirring, it began to move, to fly, and it climbed out over the wings of the Gulfstream. It moved in a wide ascending circle toward the roof of the hangar, skimming under the girders and by the posts. It levelled off. In a steady state of flight—the first and only flight in the long history of the Aereon Corporation—it circled the hangar three times, high overhead. Then it made its descent and landed at Kukon's feet.

On U.S. 1 south of Trenton, the megalopolis had come in so fast that horses trapped between motels continued to graze there. The region had two Levittowns, two turnpikes, an interstate highway, and shopping malls that were cambered to fit the curve of the earth. It was a wonderful place for a gas station—most particularly in the mile of commercial concentrate between the Pennsylvania Turnpike interchange and the Philadelphia County line. This was the site of Neshaminy Esso. The telephone rang there one hot summer night in 1971. John Fitzpatrick answered. "Neshaminy Esso," he said. "Yes. . . . Yes. . . . It was dampness inside the distributor, that's all. You're going to shoot me, right? Very well. Shoot me." He hung up, and got underneath a big Chrysler on the lift. He pounded its exhaust pipe with a rubber hammer. A shower of crud fell on his head and down the neck of his striped Esso coveralls. "I'm trying to save this muffler," he said. "If I can get this tail pipe away from the muffler, I'm in business." He pounded again, and more things fell. Fitzpatrick was in his forties—a short man, around five feet eight. His face was weather-lined, handsome, tough, and sad. There was a sense of grandeur in it, and a sense of ironic humor. His hair was dark and graying, and neatly combed. His body looked hard. He had a muscular, projecting chest. His stomach and abdomen were as flat as two pieces of sidewalk. He had the appearance of a small weight lifter, a German-shepherd owner, an old lifeguard. He was smoking the stub of a thin cigar. He pounded hard. The tail pipe would not move. Resting, he wiped his forehead with a sleeve of his coveralls, and he looked out through the service-bay doors at the Ridge Farm Motel and, beyond that, Colonel Sanders' Kentucky Fried Chicken. Although he spoke softly, he seemed to be talking to something beyond the bay, beyond the tarmac and the gasoline islands. He seemed to be talking to Colonel Sanders, whose head and shoulders— illuminated from within—were on the side of an enormous revolving chicken bucket on top of a high pole. "If you talk to any man who has been associated with lighter-than-air for any length of time, you find that they all have the big dream," Fitzpatrick said. "The dream is bringing back the big rigid airship."

The owner of the Chrysler stood to one side, as did I. The owner of the Chrysler was a mild-appearing man, not the sort who might be prepared to speak of his impatience. He shifted his weight every so often from one leg to the other. Fitzpatrick ignored him completely. On the gas-station wall was a sign that said "Yea, though I walk through the valley of the shadow of death I will fear no evil, because I am the biggest son of a bitch in the valley." Fitzpatrick also ignored the muffler and the tail pipe, and, gazing out reflectively from under the automobile, went on quietly talking to Colonel Sanders. "The story of lighter-than-air is a sad one, a story of pathos, tragedy, and mysticism," he said. "The lighter-than-air vehicle form had enormous possibilities, but they were misapplied. Airships had fantastic capabilities. They were less vulnerable to weather than any other form of transportation. They were almost immune to ordinary weather phenomena. They were much more forgiving than ordinary aircraft during conditions of instrument flight. Do you know how many people died in the Hindenburg? Thirty-six. Thirteen were passengers. Those thirteen were the only passengers who were ever lost in twenty years of commercial travel by airship, but an eyewitness announcer was there when the Hindenburg burned, and he snivelled and he cried, and the Hindenburg disaster became one of the great news events of our time. Every year, on its anniversary, pictures of the flaming Hindenburg appear in the newspapers. Since the Hindenburg, we have been through the Second World War, Korea, Vietnam: major world catastrophes. Yet every stinking year the Hindenburg appears. This is paradoxical nonsense." He picked up a hacksaw and began to saw the tail pipe. He talked sporadically as he worked, pausing between hacks to look off into the middle distance and to remember the time of the airships, and to say how odd a feeling it was to go up into the sky at a forward speed of fifteen miles per hour, and to fall in the shadow of clouds, and to rise in the heat of the sun.

"Airships had extreme range and low operating cost," he said. "They were the most economical means of air travel ever conceived. They were almost never used for what they did best. The Navy tried to use them as weapons—antisubmarine, and so on—because the Navy was in the weapons business. The future of the airships was settled on false grounds."

When the Navy settled the future of its airships, Fitzpatrick was there. He had spent the greater part of his adult life in naval aviation. He had grown up in Leavenworth, Kansas (he was an undertaker's son), and he had joined the Navy when he was seventeen, literally "to see the world." He stayed in the Navy for twenty-two years—apprentice seaman, chief petty officer, lieutenant commander. He was a pilot, and he flew heavier-than-air more than forty-five hundred hours. In 1955, he was sent to South Weymouth, Massachusetts, where he helped perform a series of tests that purported to be, in effect, a tribunal for the naval airships. Actually, it was a rite of obsolescence. Conclusions were foregone. Lighter-than-air people were known as "helium heads." They stood apart from the rest of naval aviation, phase-out-prone pariahs. The epoch of the naval airships was coming to an end. The plausible mission at South Weymouth was to test the endurance of the airships and to determine their operating envelope with respect to weather—things that had by and large been understood since the First World War. The true mission, apparently, was to collect data that could be used to kill the airships. This went on from 1954 to 1957, and the airships did not coöperate. They went out over the Atlantic on radar picket duty during blizzards that stopped all other forms of transportation—highways, railways, airports. They came back carrying ten or twelve tons of ice. Even their propeller blades were so thickly coated with ice that they were like clubs. But the airships flew, did their job, and returned, when nothing else was moving. They proved themselves anew against their supposed enemy the wind, which could not stop them with anything less than the force of a hurricane. And ultimately Lieutenant Commander Fitzpatrick, as power-plants officer, was told to work out the theoretical limits for a long cruise—intended to pass through varying climatic zones, and perhaps to match a record set by the Graf Zeppelin when it flew non-stop from Friedrichshafen to Tokyo. The Graf Zeppelin was one of the last of the great rigid airships, the behemoths of the sky. Fitzpatrick calculated the Navy cruise to accommodate the capabilities of a simple blimp called the Snowbird. Searching the literature, preparing the flight, he became more interested than he had ever thought he might be in lighter-than-air—its history, its potentiality. The Snowbird took off on March 4, 1957, and went to four continents in many weathers and extremes of temperature. There were fifteen in the crew, including Fitzpatrick. The date of departure had been set long in advance so no one could say they had picked their weather. They went through high head winds, snow, heavy icing conditions, tropical storms, deep fog, and desert sun. To conserve helium—and to avoid valving it into the atmosphere—they replaced with seawater the weight of the fuel they burned. They went down to an altitude of two hundred feet and hauled up the water in a big canvas bag. Over Morocco, six days airborne, the skipper asked

Fitzpatrick if the Snowbird could make it back across the sea. Fitzpatrick looked over his books, his gauges, and his meters—thirty-five knots, sixty-five thousand pounds of aircraft—and he said yes. In the context of the day, the Snowbird was not using much fuel. Lockheed Constellations—four-engine piston-prop airplanes—were burning two hundred gallons an hour. The Snowbird—as it maintained its equipoise between borne weight and aerostatic helium lift —was burning seven gallons an hour; and that, of course, was one of the points it was making. Its next landfall was in the Lesser Antilles, and it went up the archipelago and would have kept right on going up the east coast of the United States had the Navy not insisted that it land at Key West. The Snowbird, by then, had long since broken the Graf Zeppelin's record. It had been aloft for eleven days and had flown more than nine thousand miles—the longest unrefuelled flight, in terms of both distance and elapsed time, ever made in the earth's atmosphere. All this notwithstanding, the Navy was with full dispatch about to finish off the airships; but the Snowbird at least deserved an appropriate funeral, and that is what had been prepared at Key West. The Navy had everything from brass bands to Bull Halsey waiting for the ship when it moored.

"Bull Halsey was an extremely gentle little old man, who spoke in low, halting tones," Fitzpatrick said, and he went back to work with the hacksaw. Grit went on falling into his hair. Finally, the tail pipe dropped away, and now there remained only a small section of it, locked into the orifice of the muffler. Fitzpatrick got out a pair of gooseneck pliers that could have removed a tooth from the Statue of Liberty. He clamped these onto the stub of the tail pipe and pulled with all his strength. The section of piping did not budge. Giving up, he set the pliers down. His face was full of irony and disgust. He began to talk about a personal crisis with God that had come upon him when he was in the Navy and that he had never fully resolved. In war and in peace, he had seen tragedies in naval aviation that he could not comprehend. The disturbance he felt had not decreased in the years after the Second World War but had, if anything, intensified. "I believe in a God. I believe in a Creation," he said. "I can prove the existence of God without any trouble at all. You can see by looking around you that this is an orderly world we live in." He glanced at the environment—Hess, Sunoco, Texaco, the Trail Blazer Diner, Krispy Kreme Doughnuts, Lincoln Drive-In Theatre, Gino's. "There can be no order without a God," he went on. "Intelligence is God. The order of the planets follows no Mendelian laws, no Darwinian theories. Call it God if you will—a scheme that knows no limits of size, of place, of power. You will never find a serious mathematician or physicist who is an atheist. In the study of higher mathematics you can sense it. You are finally hearing the language that man and God can understand. Scientists and engineers are the true priests. These are the men who have given us the tools to dominate nature—to reach out to the stars. Practically everything that is noble in man has been given us by these men of mathematics and science. The church has failed totally." He broke off, was silent for a time, then drifted back to his Navy experiences and the deaths that went past his understanding— "casualties of war, casualties of the air, accidents on flight decks: seeing a man get up and walk ten feet before he discovered that the back of his head was missing." He said he wondered if God marked each sparrow's fall. He wondered if man had created God. He wondered even more when his son, who was operated on because he had a double nerve in his spine, died. After the flight of the Snowbird, he said, he was transferred to a Navy unit at McGuire Air Force Base. In the unit, one man he had particularly admired was a chief petty officer who was killed while returning to the base from leave. The chief's wife was killed, too, and a son was crippled. Fitzpatrick said he had sorted through the man's effects in bitterness and anger. "There he was, a good man, dead in the middle of the morning, doing everything right—killed by a drunken girl who hit him head on." Fitzpatrick then tried to set up a memorial service for the chief in the McGuire chapel, with the entire naval unit attending, but an Air Force priest, who happened to be the supreme cleric on the base, pointed out that some of the men were Protestants, some Catholics, some Jews, and the priest would not allow an interdenominational service in the base chapel. So Fitzpatrick scheduled a service in a hangar and asked the Navy to send in a chaplain from somewhere else. The man who came was Monroe Drew, U.S.N.R., the minister of the Fourth Presbyterian Church of Trenton. "I had little use for chaplains," Fitzpatrick said. "But this one preached a sermon that was not pap. The essence of what he said was 'We don't know why these things happen.' I liked him for it. In the end, he would disillusion, disappoint, and enrage me. This is an understatement—I do not like Monroe Drew."

The mind of Monroe Drew at around the time of the

chief's funeral, in 1959, happened to be aflame with practical visions of the return of the big rigid airships. His conversations were intense with plans for fresh awakenings and novel departures in the airship field, and to implement his ideas he had actually been raising money. Fitzpatrick was a godsend to him, or so he thought—a man anxious to vindicate the concept of lighter-than-air, an engineer experienced in airships. He could design and build Drew's dreams. Fitzpatrick, attracted, retired from the Navy. He and Drew formed the Aereon Corporation.

Now Fitzpatrick got out an air-powered jackhammer and went at the Chrysler with ear-ringing blasts. The hammer slipped. Its prong went through one side of the muffler and out the other, destroying the muffler in a single penetrating thrust. Fitzpatrick tried to remove the jackhammer, but it was stuck solid. He said nothing, showed no diminution of composure and not even much interest. "I have an inherent distrust of people who think big," he said. The telephone rang a few times. He went to it and picked it up. "Neshaminy Esso," he said. "Yes. . . . Yes. . . . No. What you are saying is that the spirit of the law is sometimes different than the letter of the law. I have no way of telling if your complaint is legitimate."

THE DELTOID AEREONS were not conceived until 1966. Before that, an aircraft of a different shape was slowly put together, and it was an authentic wonder. The company was less security-minded in those early years—no shrouds, no Sheetrock boxes—and people who happened by the hangar at Mercer County Airport saw something in plain sight that they were unlikely to forget. Poised in the air over tricycle landing gear was a Siamese-triplet dirigible: three rigid cylindroid hulls, each eighty feet long, parallel to one another, welded together at their longitudinal equators, and covered with a milk-white skin that was dazzling even in the hangar's gloom. It was as if the Akron, the Macon, and the Los Angeles—the last of the Navy's zeppelins—had been scaled down to eighty feet, placed side by side, and united as a single aerial catamaran. If the thing had been created somewhere else, it might have received more attention than it did, but this airport was only five miles from the center of Trenton, and in Trenton civic humility was the most becoming aspect of metropolitan life. People in Trenton seemed to have newts' eyelids that had lowered over the suggestion that anything in Trenton could be interesting. One autumn Sunday in 1962, the filmmaker Tom Spain, who lived elsewhere in Mercer County, was driving around with a friend and went by the airport. In idle curiosity, he stopped his car, got out, and peered through a hangar window. He saw the triple hull. "Jesus!" he said. "What in the name of God do you suppose that is?" He thought it was extremely beautiful—a sculpture of converging lines and myriad triangles. In days that followed, he asked around Trenton: Could anyone tell him what he had seen? Oh, everyone knew what that was. "That's Monroe Drew's airship." The reply was as flat as it might have been had Spain asked why there was a gold dome on a large building near the heart of town. Spain, at the time, was filming television commercials depicting Marlboro Country in wild pockets of the megalopolis outside Trenton. No one in Trenton thought that was remarkable, either.

Monroe Drew's range went some distance beyond the pulpit of the Fourth Presbyterian Church and the chaplaincy at the Naval Reserve Training Center on Lamberton Street. He was, for example, the author and originator of Teleprayer in Trenton. You dialled LYric 9-4574 at any time of the day or night and you got thirty seconds of recorded spiritual assistance, in prayer form. The prayers changed daily. They were sixty words long, and the Reverend Mr. Drew composed them carefully, running, as he put it, "the general gamut of human needs." Sometimes,

Drew said, the Teleprayer attracted fifteen thousand calls in a single day. Drew had a deep conviction that the extraordinary novelties of contemporary technology ought to be used in all possible ways to shape the world in the way of Christ, and he felt that this potentiality was being absurdly neglected. Teleprayer's fifteen thousand calls in a day were a small example of what he meant. He meant that Martin Luther had used the printing press to effectuate the Protestant Reformation. Power of that kind, freshly generated in the technological outbreaks of the twentieth century, was waiting for a new Luther. Motion pictures were the most obvious medium. Drew had made religious films during eight years he had spent on active duty as a chaplain in the Navy, and he had later been a high functionary within the audio-visual precincts of the Presbyterian Church in the U.S.A. In both situations, he had become frustrated and somewhat embittered by the indifference he reaped and by the absence of large-scale distribution of his films. In Trenton, in 1958, a son of his came home from grade school one day with a copy of *My Weekly Reader* that contained an article about the phasing out of the naval airships. The boy asked how the airships worked, and Drew—who was not a pilot of any kind, but who had been around naval aviation a great deal—outlined the elementary principles of lighter-than-air flight. That evening, when he went to his study to compose a Teleprayer, he developed writer's block. He doodled for an hour and was unable to line up so much as three words in a row. Eventually, his thoughts were diverted toward dirigibles. Why not use exhaust gases to heat the helium and increase lift? He forgot the prayer and sketched a zeppelin. Why not build lighter-than-air vehicles big enough to carry awesome loads? Houses, ship screws, industrial generators. Why not use lighter-than-air to revolutionize missionary aviation? Why not create a Faith Fleet, a Christian Freight Line marked with the insignia of the National Council of Churches, to carry food, goods, and Bibles to people in what the church called the opportunity countries? Fifty transformers to the Voltaic Republic, a hundred thousand Bibles to Nigeria, a million peaches to the Haut-Katanga. Why not bring the world's underdeveloped nations into the transportational forefront of the twentieth century in a single leap, by eliminating the need for roads, railroads, tunnels, bridges, airports, storage facilities, and prepared harbors? Enormous warehouses in the sky would move from place to place, landing lightly on grass fields. Why not operate the Christian Freight Line in the United States as well? This prodigious advance in the airfreight business would result in the acquisition of millions of dollars, with which a foundation, not unlike the Ford Foundation, could ultimately be established. Why not allocate some of the foundation's funds to the making of films and other audio-visual devices communicating the word of Christ in modern terms? And why not spread these films and audio-visual devices throughout the world in the airborne hulls of the Faith Fleet? Why not? Why not? With contemporary materials, with modified designs, why not bring back the rigid airships?

In time, Drew reworked these thoughts into a preoccupation, and then converted that into an obsession. He sought financial support in various sections of Trenton—from the treasurer of the Fourth Presbyterian Church, from parishioners, from merchants, from industrialists. He expanded his quest beyond the city. To virtually everyone, he said, "How much can you afford to lose on a project that might be of truly great importance to the world?" Drew sensed that he had been given a mission. Earnest, humble, blue-eyed—his hair combed straight back, his gaze steady— he spoke not with soapbox intensity but softly, with an almost mournful sincerity; and he had the forensic grace to transmit to others a share of his enthusiasm. Seed money resulted. All he lacked was a technologist in lighter-than-air. The telephone rang. The United States Navy ordered him to perform a funeral at McGuire Air Force Base, where, on arrival, he was to report to Lieutenant Commander John Fitzpatrick.

Drew organized the Aereon Corporation around John Fitzpatrick, and eventually made him president. They worked together for seven years. "John can worry about strength factors, shear forces, sharp gusts, and thunderheads," Drew said. "I'm more free to dream." He assembled a board of directors—a hospital administrator, a management consultant, a patent attorney, among others. He asked the session of the Fourth Presbyterian Church to allow him to spend half his time on the Aereon project in the interest of missionary aviation and with an eye to future profits for the treasury of the church. The session approved. A woman schoolteacher who was an officer of the church invested fifteen hundred dollars in Aereon. Outsiders came in with bigger sums. Lefty Klein ("a Jewish gentleman"), owner of a vinyl-fabric firm in Trenton, invested five thousand. A stockbroker who owned his own P-51 put in ten thousand. Richard Hoerner, of the Hoerner-

Waldorf box company, put in forty-five thousand. Laird Simons, the kid-leather king, began an investment that would eventually reach a hundred and fifty thousand dollars. Promoting Aereon, Drew frankly stated that his ultimate goals were to advance the cause of the church. "I know what I want to do," he said. "I know what I want to do right here in Trenton. But it's going to take millions to do it. The result will be a breakthrough in the economics of transportation that will sweep across the world, at the same time serving the needs of church evangelization—a natural tie-in. The two dovetail." He went to the Presbyterian Foundation seeking a list of their major contributors but was reverentially turned down. He hired a Japanese photographer to record the developing history of the corporation. He hired Vice Admiral Charles Rosendahl, U.S.N. (Ret.), the greatest living figure from the era of the naval airships, as general consultant. He engaged a Washington representative to establish close communication between Aereon and the patent office and other parts of the federal government. Looking into the existing situation in missionary aviation, he found it farcical. "The Missionary Aviation Fellowship is controlled by people who believe that God had a typewriter," he reported. "They take Scripture literally. They ask you, 'Brother, are you born again?' And Heaven help you if you answer no. They have one Waco biplane. 'Is that all you have?' I asked them, and they said, 'That's all the Lord provided.' " Drew did not tell them that the Lord was about to provide the Christian Freight Line. He investigated the Roman Catholics to see what they were doing in missionary aviation, and he turned up a few pilot priests and some more bi-planes. That was all. So he widened his vision of the Faith Fleet to include ecumenical Aereons that would transport Catholics, Congregationalists, Baptists, Unitarians, Lutherans, Presbyterians, Methodists, and their goods to far-flung missions for five cents a ton-mile versus the dollar they were spending at the time.

Fitzpatrick, for his part, felt pressures rising. The company was heavy with consultants, representatives, photographers, attorneys, and other overhead. Before the money ran out, he was expected to design and to supervise the construction of an airship that would change the world. Fitzpatrick was the right sort of man to make such an attempt, for he liked to go from A to B without inventing letters in between. Ignoring such delays as blueprints, small models, and wind tunnels, he built Monroe Drew's airship off the cuff. "Has it occurred to you that a wind tunnel is simply an admission of defeat?" he said. "Certainly it is. A wind tunnel is simply a sophisticated version of the old cut-and-try method. The *existence* of wind tunnels proves that. Mathematics alone is not enough. Designers use all the information they have. They build a model. They stick it in a wind tunnel and hope that they've got all the basic data correct, that the arithmetic is good, that their basic judgment is right. Sometimes it is. Sometimes it isn't. In the end, they have to resort to empirical means. What is true for a model is not necessarily true for a full-scale aircraft. Think of a drop of water hanging for hours off the end of a faucet. There is a decided limit to the size of that drop. You couldn't take that drop of water as a true indicator as to how a ton of water would act. I doubt if you've ever seen a ton of water hang off the end of *anything* for very long."

It happens that a cubic foot of helium will lift about one ounce. The ratio between the surface area and the volume of a cylinder is not constant: the smaller the cylinder the more surface area, and thus the more weight, there is in proportion to volume. Fill with helium a small, rigid, cloth-covered aluminum-framework cylinder, and nothing happens. It is too heavy. Fill a large cylinder built the same way, and it rises into the air. How large? "Eighty feet," said Fitzpatrick. "Anything smaller than that is a parade float." He made the triple-hulled Aereon eighty feet long—the smallest model that could test all aspects of the concept at once. The concept was to create a hybrid of the airplane and the rigid airship—something with the buoyancy of an airship, the aerodynamics of an airplane. The three combined hulls, in their considerable breadth, would serve as an airfoil, however coarse. He used the Navy's Class C airship form, because its profile produced the least drag of all airship hulls. He called the ship a "flying fuselage." It weighed a ton and a half, and it carried its fifty thousand cubic feet of helium in plastic bags. Its gearbox was the rear end of a Chevrolet. Its engine was a seven-hundred-dollar four-cylinder McCullough, of a type used on drone aircraft—cheap to buy, cheap to run, sufficient for the job, Fitzpatrick said, because the helium would assist it. The helium would considerably prolong the engine's life as well. Unlike an airplane engine, it would not have to be gunned to death at full bore on takeoff. "For an airplane, even a little sixty-two-horsepower turbine sells for about twelve thousand dollars," Fitzpatrick said. "Figure that out on a per-pound basis. You're better off eating the money." His ultimate goal, after all, was to build a version of Monroe Drew's airship that would move long distances through the sky cheaply, carrying two hundred tons of cargo. For propulsion and control, he mounted a helicopter rotor vertically on the stern of the middle hull. It could be used for both braking and steering at dead-slow speeds, when other control surfaces—elevons, rudders —would be ineffective. By regulating the temperature of the helium, the ship, while on the ground, could be heavier-than-air and thus far more stable. Heating the helium would add considerably to the tonnage of liftable cargo. Because the ship could fly aerodynamically, it would require no mooring mast, and it could land when it wanted to. "Airships used to kite and balloon while men were trying to tie them down," Fitzpatrick said. "The big rigids cruised for hours waiting for right conditions for landing. We're independent of that. What we have here is a revolutionary aircraft."

The triple hull took years to complete. On Admiral Rosendahl's recommendation, the Reverend Mr. Drew hired the airship rigger Everett Linkenhoker to do the actual labor, and Linkenhoker built the frame, weld by weld, while Fitzpatrick, at his drafting board, tried to stay a day or so ahead, and Drew, at his desks—in the church, in the hangar—played the light of his thoughts across future decades. It was up to him to generate money, and he did so with infectious images of thousand-foot automated Aereons moving in connected trains through the lower atmosphere. An Aereon would one day hover above an orchard, pick up the entire crop, and deliver it by winch to waiting markets, he said. An Aereon would call at Moline, Illinois, hoist up a hundred tractors, and deliver them to the Valley of the Rhône. Everything would be organized in containers—fuel, cargo, even crews—to maximize efficiency. Containerfuls of commuters could be winched into the Aereons, moved over the megalopolis, and deposited in city cores. "I envision new cities built around this type of aircraft," Drew said. "It will render obsolete road and rail connections. Today's big helicopters can lift nine tons. We'll stick nine tons in a back corner. At high altitudes, where air is thin, airplane payloads go down as landing speed goes up. Five thousand feet is about as high as an airfield can be. The Aereons will be just as useful at ten thousand as at sea level. They are not penalized by altitude. The airship coöperates with nature. The airplane fights nature."

Fitzpatrick formed the opinion that Drew was putting out "enough propaganda to gag a maggot." He said, "I don't care how owlish you look, how convincing you sound, this is just yak yak yak until you do it. You have no product until you have done it. All you have is pie in the sky." But Fitzpatrick was a believer, too. "We'll never compete with jets in carrying Paris furs, cut flowers, gems, corpses, and refrigerated serums," he once said. "But these are the exotic fringes of freight. The first time we lift a bulldozer, it will be very impressive."

When the eighty-foot Aereon was ready to roll out of the hangar, it was a little heavier than intended, the rotor-prop required shortening because it created such grossly excessive vibration that it might well have shaken the airframe to pieces, and the power of the engine was, at best, marginal. None of these problems bothered Fitzpatrick. Experimental aircraft always have problems—items on the agenda of cut-and-try. "Before the first flight, we will have made speed runs and zoomed into the air for short distances to permit the human occupants of the thing to become familiar with it," Fitzpatrick said.

"It is such an unusual-looking aircraft," said Drew. "We expect a pretty definite reaction from the population."

"It will look like an aluminum overcast," said Fitzpatrick.

On April 15, 1966, in a fifteen-knot crosswind, Monroe Drew's airship made a taxi run on a Mercer County runway, failed to slow down, tried to turn at the end, tilted on two wheels, and suddenly became a sail in the wind. "We can bring the nose of the ship down into an angle of attack that is negative, so that strong winds, rather than making the ship a kite, will simply cause it to hug the ground tighter," Fitzpatrick had said. "We have introduced tricycle-type landing gear of extremely wide tread and extremely long wheelbase to increase the stability of the ship on the ground. This will be the first time that an airship has ever had three-axis control." In the Plexiglas-bubble cockpit at the forward end of the central hull were two airplane mechanics, one of whom was also a small-airplane flight instructor. These were the test pilots. Both developed panic. One jumped out the hatch, even though the drop was eighteen feet. The wind then blew the airship over—flat on its back. The other pilot, who was still in the cockpit, jumped straight down, smashing through the Plexiglas and falling out onto the ground. The weight of the pilots was critical—three hundred and fifty pounds of ballast gone. The wind overturned the airship again. Damage done by the wind was considerable, but embarrassment caused even more. Drew and Fitzpatrick decided to get the wreckage instantly out of sight, and the aluminum overcast was virtually bulldozed back into the hangar, arriving more or less in flakes. One of its nose cones is now in the collection of the Lighter-than-Air Society, in Akron, Ohio.

The end of Monroe Drew's airship appeared to be the end of the Aereon Corporation. The company was six years and seven months old, and its fixed assets were one crumpled dirigible. Optimism was replaced, in part, by friction, and friction, often enough, by combustion. To Drew, Fitzpatrick's confident empiricism now seemed a brightly lighted shortcut to nowhere. Fitzpatrick might be a genius, Drew said, "but he's an arrogant genius—one of the most arrogant men I have ever known."

"I look down on—generally speaking—ministers," Fitzpatrick decided. "They get between man and God. Drew has around him a charisma. He can quickly pick up the patois of almost any profession, without talking intelligently. He believes he is meant for greatness. He believes there are forces that thwart him on the threshold of success. There certainly are. Long ago, Drew rubbed elbows with greatness. He sang on the radio as a boy."

Drew was, indeed, a boy celebrity. At the age of fifteen, he had become a professional bass-baritone, singing florid segments of "The Desert Song" and "The Vagabond King" on the Columbia–Don Lee Network. He was a scion of Lawrence Tibbett. His expectations were high. He was part of a San Francisco family that had once built square-riggers for the China tea trade. His father was a Presbyterian minister. Fitzpatrick, after absorbing all this in fragments over the years, had become fond of telling the story of Drew's father's death, as narrated to him one day in the hangar by Drew. When his father was dying, Drew was in seminary and was about to become a minister himself. The old minister called Drew to his bedside for a parting word of advice. "Never trust another minister," said the father to the son, and died.

Drew spoke less anecdotally about Fitzpatrick. "John's Achilles' heel is that he has an inordinate amount of pride—a really stupid pride," Drew would observe. "He didn't lie to me, but he didn't tell me all the facts about the error he had made—about the weight of the ship. He is a very good engineer, though. A self-taught man. He once came very close to getting his degree."

A management consultant was not needed to see that Aereon had entered a state of Cheyne-Stokes respiration. On no known economic indicator would the company have scored as high as zero. But it was not about to die. The term "helium head" had not developed casually. These men were not just building any aircraft. They were bringing back the rigid airships. They had support—mostly moral, to some extent financial—in all the side streets and shore cottages where the outphased, bypassed, disenfranchised men of the naval airships had settled down to disappear. In their separate ways, Fitzpatrick and Drew were both helium heads, and zero was not a low enough level at which to decide to quit. Maybe the triple hull was not the best configuration after all. It was something of an airfoil, yes, but maybe there was still too much friction drag. There might be a new and better solution. But what? Where? How to find it? Drew had repeatedly prayed for the milk-white Aereon. The wind had answered. Now where could they turn?

In Valley Forge, Pennsylvania, the General Electric Space Center maintained a computer so august that one did not actually approach it in person but communicated with it by teletype. Drew and Fitzpatrick decided to commit the future of their project to the decisions of this machine. To do so, they had to make a list of criteria, a statement of what they wanted to accomplish, and to help prepare the list they hired Jürgen Bock, a physicist from the Max Planck Institute in Heidelberg and more recently from the U.S. military proving ground in Aberdeen, Maryland, who joined Aereon full time. Bock spoke in English, thought in German, and dreamed in a language no human being had ever heard. In the Mercer County hangar, therefore, communication was not perfect, but it had loft. At length, the list was complete and was teletyped by Bock and Fitzpatrick to the computer. They wanted an airplane-airship, they said, and they wanted to find the optimum configuration for enclosing maximum volume without too much penalty of drag. In numerical language, they went through the whole vision of grass airfields and flying warehouses, titanic cargoes, dollar efficiencies, cheap power plants, bagged helium, aerodynamic lift, long range, great endurance, and specifications for control of the airflow around the hull. In all, twenty items—twenty matrices—were sent to the computer. A response came back at once. The computer said, "Remove three matrices. I can only handle seventeen." Conversation followed in Mercer County. The three least essential items were removed. Silence followed in Valley Forge, while the computer ingested the seventeen criteria and splayed them into hundreds of thousands of crosshatching possibilities. It selected the best of these, shaped them mathematically, and sent back to Mercer County Airport the precise configuration of the 7, the 26—the deltoid Aereons.

I N   T H E   P U M P K I N - S E E D   configuration, helium would be ineffective at any length that was much under a hundred and fifty feet. So helium was not included in the tests of the 7 and the 26. They would have to get up there on their own—without wings, without lifting gas—and ratify the computer that had created them. Then the way would open to the behemoths. The 7 was tested for the fourteenth, and last, time on September 2, 1970, near Princeton. Monroe Drew was not invited. He was in Trenton—eleven miles, and many light-years, away. John Fitzpatrick was in Neshaminy, pumping Esso. Drew and Fitzpatrick by now were like old Russian premiers. They had been there once, but no one quite remembered them. When Miller had become president of the company in 1967, he replaced Drew's overt and provocative optimism with a hair shirt of caution and secrecy. Something called the Technical Advisory Group replaced Fitzpatrick. The image of Aereon flew by night, and Miller dedicated himself to building a different impression. He found his consultants, in the main, in the Department of Aerospace and Mechanical Sciences of Princeton University. Putman, the aerodynamicist, was one of these, as was Kukon, and Olcott had been trained there. The evening of September 2nd was heavy with the heat of the day. Air was moving, but barely. Everyone wore a sports shirt but Olcott, who was dressed in a plain brown suit and did not remove his jacket or tie. The advisory group formed a circle around the 7, at the head of the runway. "Today's outing must be a success or a complete failure," Putman said. "We are here to make a flight. And to get a flight—to get out of ground effect and make a circuit of the field—we are willing to risk totalling the 7."

To salvage the aircraft, Kukon had worked for a week, steadily and far into the night, in a shop near Princeton. The work had been done so skillfully that the others had to think to remember which side had been crushed by Putman's car. The orange silk glistened with fresh, clear butyrate dope. Kukon showed no trace of concern for the 7's immediate future. It was part of his routine to build with uncompromising care devices that might be destroyed in an instant. He did say, however, "If I can't get above six feet, I may just set it back down in the grass."

Olcott said to Kukon, "In trying to get out of ground effect, try not to do it too quickly. Let the vehicle go out at a shallow angle. Don't raise the nose so much that you start to lose air speed." Within a few days—everything going well—Olcott was expected to get into Aereon 26 and fly it. He had been flying it for the last couple of days on a computer. Now he wanted to see if the model would reflect or contradict what he had learned. He wanted to see how it would behave when it got out of ground effect, among other things, and what sort of roll angle would be best for a turn, what the diameter of the turn would be, and how much elevator deflection would be needed to fly at a given altitude as a function of center-of-gravity location. He said, "Try one control input at a time, John. First try pitch excursions. I don't want anything gross. Just small things. So we can see what's happening. Try coördinated rolls. I think the vehicle may have N delta A, adverse yaw, a little more than predicted, and the only way to take care of that would be with rudder. So you may have a problem with just-aileron rolls. I think where this vehicle can get in trouble is that it can become grossly uncoördinated. If you let the sideslip build up quite a little, I think you'll find that it becomes unmanageable. In the 26, I'm going to try very hard to keep everything coördinated, to keep that ball right in the center. The simulations indicate that life remains manageable when the ball is in the center, but if the ball gets way out, if the sideslip angle builds up, life can become very difficult."

"Gentlemen, I think we ought to do it," Putman said.

"O.K.," said Kukon.

The engine started—small, rear-mounted, earsplitting—and the 7 taxied around, limbering up. Everybody got into Putman's convertible. As before, Putman would drive in close formation with the aircraft, but when it went into the air he was to stop. Kukon sat on the convertible's trunk, his feet dangling into the rear seat, his Logictrol transmitter in his lap. Miller was beside him with his Super

8. Olcott was in front. Kukon advanced the throttle. The aerobody and the automobile raced side by side. Putman called out, "Twenty! Twenty-five! Thirty! Thirty-two!" At thirty-seven miles per hour, the 7 took off.

The wreckage this time was total and irreparable. Small pieces of Styrofoam, balsa wood, piano wire, and orange silk were scattered across a dirt road and into a cornfield. No remark showing disappointment or dismay was made by anyone. They all got out of Putman's car, stood in the middle of the orange and white debris, and talked about what had happened. Miller's movie camera, clicking as if it were packed with tree frogs, committed each fragment to film. Olcott looked reflectively at the bits and the pieces, the Styrofoam spread out like snow in the late, slanting light. He spoke at length with Kukon, and as he did so he flew his own hand, in various pitch attitudes and aileron rolls, in the air before him. Now and again, he grinned. The 7 had at least *flown* to its final destination. It had flown high and, for a while, straight.

Putman said, "We haven't seen anything shockingly unusual on this outing. I don't think the 7 had stability problems. I think it had control-effectiveness problems and trim problems. Any power problems, John?"

"No. Here's the story, and it's very strange," Kukon said, speaking rapidly and supporting his words with gestures. "The aerobody lifted off very gradually. Then it seemed to settle. I gave it more up-elevator. It sort of got a little soggy. I gave it more up-elevator. It climbed a little—to three or four feet—and I gave it more up-elevator, which was probably two-thirds of what I had. I held that elevator, and the aerobody very gradually started to settle. At that point, I had to make a decision. Was I going to throttle back and set it down, or go all the way? I decided to go all the way. I throttled back just a little bit, though—three notches. And when I throttled back, the whole thing levelled off and climbed. It surprised the hell out of me. When I saw that, I gave it full throttle and full up-elevator. It kept on climbing. I felt a very strong trim change when it moved out of ground effect."

"In ground effect you need a lot less elevon?" Putman said.

"Exactly."

"The proximity of the ground gave you a more positive angle of attack?" Olcott said.

"Right."

"O.K.," said Putman. "This would indicate to me that we need a more nose-up elevator after we get out of ground effect. With the more nose-up elevator, if the separation and scale effects are severe, your aileron effectiveness will accordingly go to pot. So you may have a lot less aileron effectiveness out of ground effect *not* due to the fact that you are out of ground effect but due to the fact that you need more elevator."

"Close to the ground, everything was sensitive," Kukon said. "I had a lot of power. I could put it where I wanted. Everything was terrific. Even the rudders were more effective. Once I got a little higher, though, it was as if I was trying to hold a long rubber band that was hooked into the controls."

"There's a distributed lift effect," Putman said. "It might be less severe on a full-scale machine. I don't think it is something that we can do anything about at this point, but ground effect is the major unknown area that is going to give us problems."

"After it had climbed out, I tried the ailerons a little bit, and they were responsive to low roll angles," Kukon went on. "Everything seemed O.K. But then the aerobody rolled sharply to the left. I gave it some right aileron. This was at fourteen feet of altitude. I gave it some more right aileron, and it really wasn't responding. Then I gave it some rudder, and, boy, did it respond! That was the first time I used rudder, and it just zipped right around and straightened itself out. I neutralized the ailerons and neutralized the rudder. That's when the aerobody went up to forty feet. Flying straight. Everything fine. Until I tried to turn around. I rolled it over with a little bit of aileron. It held there pretty well, but then it started to slip off to the left. I straightened out the ailerons. Nothing happened. The amount of roll that I was allowed on the whole flight was essentially zero. I mean, just a little bit and it was already too much. I gave it a little more right aileron. Nothing happened. I gave it full right aileron. Nothing happened. I gave it full right rudder, and it wasn't enough—full right and full right! There was no way I could get out of it. It was over too far. It spiralled down."

"That seems very close to what we learned in the computer simulations," Olcott said.

"If you ever get to that point, there's no way out," Kukon said.

"The thing is not to get to that point," said Olcott. "With this aircraft we have to be very careful that we don't let things get too far out of trim."

"I agree."

"Because by the time we got that second rudder input—"

"It would be too late."

"—we would have gone too far. Do you have any notion what might happen if we did not use the ailerons but just the rudders?"

"I think the vehicle would be uncontrollable," Kukon said. "Because the motion I got out of pure rudder was a wild thing. The whole flight was very difficult. I was working pretty hard, I thought, even during the straight portion. One puff, one gust, one very small disturbance, and I'd have lost the whole ballgame right there. That one time I put in full rudder was almost a panic control."

"John, thank you," Olcott said. "Thank you for that flight."

"I'm sorry I couldn't get it a little better around for you, but that was the best I could do," said Kukon.

"Flying the 26, we'll have a lot more information at our command than you had flying the 7—a lot more abort opportunities," said Olcott. "The whole philosophy of the 26 is that we're not going into any unknown area quickly. We'll go in small, manageable steps—so we can always back out. I'm not going to force the 26 to do anything it doesn't want to do. I don't intend to force it into the air. It has to fly into the air. Meanwhile, the 7 has told us a heck of a lot. The 7 supports the simulator study. There is no glaring conflict. It indicates that N delta A is a little higher than predicted and that L delta A might be just a little less than predicted, but we've got a consistent set of data." Olcott unbuttoned his jacket. A lock of hair had fallen across his forehead. He paused a moment and looked down into the wreckage, but he was not really seeing it. "That flight was worth its weight in gold, John," he said finally. "We have a way of approaching the 26. We've identified a risk. Now we want to say, 'If we can keep it small, we can handle it.' I think we ought to go to NAFEC. I think we ought to go into an attempted lift-off Friday morning."

O N S A T U R D A Y E V E N I N G , September 5, 1970,
John Olcott and his wife gave a dinner party at
their home, in Basking Ridge, thirty miles north of Princeton. Beforehand, Olcott dialled the Aereon Corporation's
number and spoke with an answering service. He gave his
name, and waited while it was checked against a list. Rain
and wind had delayed the first attempt to fly the 26. Now
the message was "Good weather tomorrow morning. The
winds will be south-southwest three to six knots at 6 A.M.,
building up to six to ten knots by 8 A.M." Aereon had re-
served dawn to eight Sunday on the big runway at
NAFEC. The guests came. Olcott prepared drinks. He did
not want to be a killjoy, so he privately mixed himself
water on the rocks with an olive. He revealed nothing of
his plans until after dinner, when he said he was sorry but
he had to work early the next morning; and abruptly he
went to bed. He got up at three, and although he was on

his way to a remote and all but deserted airfield, he dressed in a blue button-down shirt, a dark-blue narrow tie, gray flannel slacks, and a blue-white-and-yellow madras jacket. By three-thirty, he was moving south in his Karmann Ghia—Interstate 287, the New Jersey Turnpike, the Garden State Parkway. At seventy miles per hour, glued to the banked turns, he was going about as fast as Aereon 26 would ever go or was ever meant to go. He moved under stars through mile after mile of dark-corridor forest on the eastern perimeter of the Pine Barrens. He crossed over the Mullica River, shot through the vacant streets of Pomona, and went on through a countryside of open fields and stands of pine to a set of gates in the high chain-link and barbed-wire fencing that surrounded the flat immensity of NAFEC. Inside, he drove on for another mile or so, until windowless walls seventy feet high loomed up black before him against the barely graying sky. He parked the Karmann Ghia at the corner of Lindbergh Drive and Firehouse Lane. Carrying his flying boots and his test pilot's note pad—his jump suit over one arm—he walked toward the big building. He went through a small door in an acre of wall, like a mouse going home.

More cars came into NAFEC—small auroras moving in through the darkness and blinking out near the big hangar: William Miller from Princeton, William Putman from Staten Island, Charles Mills from Toms River. Mills, a German teacher in a high school on the outskirts of Trenton, had once been Air Operations Officer at Lakehurst Naval Air Station. He was now an Aereon consultant. He had been a celebrated pilot of naval airships. When the airships, just before the Navy abandoned them, gave ironic proof of themselves flying out of South Weymouth, Massachusetts, in the middle nineteen-fifties, it was Mills who piloted the most dangerous flights. He rigged up closed-circuit television so he could watch ice coating up on his Z.P.G.s—big ships, three hundred and forty-three feet long —until he had as much as five tons of it weighing down his leading edges, his propellers, his control wires, and his windshield. Then he would hunt for the worst of a storm. Mills had the feel of the airships, of the wind and the weather. His responses were quick, and he knew how to use them. He had an athlete's sense of anticipation. An airship moved cyclically in pitch and yaw. Anticipated lead time was the middle of the art. When the nose was going down, the moment was right for some down elevator to check the up cycle that was coming later on. Pilots who did not feel this could let their ships stand on end or slide, giddily, sideways. Mills liked to take the Z.P.G.s out onto the triangle of runways at South Weymouth and slide around on his landing gear with the precision of a figure skater, upwind, downwind, crosswind—strange exercise, a waltzing cow. Watching all this from an engineering office, Lieutenant Commander John Fitzpatrick, who did not overestimate other men's abilities, came to regard Mills as "a master of lighter-than-air flying." To show just how much strain a dirigible could withstand, Mills deliberately flew one carrying eight thousand pounds of ice into a front of warm air, and somehow—through touch, verve, whatever—emerged safely on the far side of an aerial avalanche. For that flight, he was given the Harmon Trophy. Mills now opened the small door in the big wall and stepped into the brightly lighted interior of the

NAFEC hangar, where single-engine, twin-engine, and four-engine aircraft were spaced out on fifty thousand square feet of smooth concrete, and where reciprocating engines, out of their nacelles, had been set up on mounts and looked like big women in curlers.

"Hello, Charlie."

"Hello, Charlie. Did you have any trouble getting up this morning?"

"Negative."

Aereon 26 was in the central foreground, bright orange and lustrous, fat, sleek, and implausible, with its black stripes, its black markings, its nose-mounted Pitot boom pricking forward a full six feet. This boom, from which a little stub-winged airspeed indicator hung like a model rocket, was about the only added feature that made the 26 different from the 7—other than, of course, its scale. Everett Linkenhoker was crawling around inside the 26, completing its final preflight checkout. He had been there most of the night.

Mills put his head in through the hatch—a squarish head, with short hair, blue eyes, reading glasses hanging from a cord around his neck. "Hello, Link."

"Charlie." Linkenhoker spoke around a toothpick, without really looking up. His hands were on the airframe. His eyes were moving from weld to weld. He was a short man, heavy in the cheeks, heavy in the middle, a quiet, contemplative man, inventive within his realm. He had light-blue eyes under bifocal lenses, and sandy-blond hair that had gone partly to strings. When Mills had been Air Operations Officer at Lakehurst, Linkenhoker had been a petty officer there, rigging airships.

In volume, Aereon 26 was twenty times as capacious as light airplanes of the same length. At eleven hundred and forty pounds, it weighed about half as much. If its structure had been formed from sheet-metal jigs, the way modern airplanes are built, Linkenhoker would have been able to move around quite easily inside it, but Aereon had had no sheet-metal-working equipment nor the money to buy it. What the company did have was Linkenhoker, a Heliarc welder, and in building the airframe he had used neither a bolt nor a rivet. The 26 had a totally welded tubular structure, consisting of many hundreds of slender aluminum rods compiled in intricate rhomboids, trapezoids, triangles. Drenched now in filtered orange light, the interior of the 26 seemed to have been composed rather than engineered. Or it might have been some prize-winner's discovery in organic chemistry—a novel molecule magnified eight hundred billion times. These aluminum tubes had been salvaged by Linkenhoker from the wreckage of the triple-hulled Aereon, as had most of the 26's instruments: the cylinder-head temperature meter, the free-air temperature meter, the altimeter, the artificial horizon. Piper Aircraft asked thirty-two hundred dollars for a big Pitot boom of the type Aereon wanted (with a yaw instrument, an angle-of-attack instrument), and that was beyond Aereon's means, so Linkenhoker borrowed a set of plans and made the boom himself. He built the fuel tank—ten gallons—out of new sheet aluminum, and he mounted it to the aluminum rods in the exact center of gravity. To make landing gear, he bought at an auto store some half-inch-diameter bungee shock cords—the things that keep suitcases from falling off roof racks—and he cut them into pieces eleven and a half inches long, bunching up six for each main gear and four for the nose gear, attaching them to landing-gear tripods with aircraft wire and U-shackles. He foraged at small airports until he found an appropriate wreck and from it took the 26's brakes. He found the plastic cockpit canopy on a decaying glider. One day there was an odd and tragic accident at Red Lion Airport. A mechanic flew off in a Cessna to another airport to pick up a case of oil. It was a bumpy day. The Cessna had dual seats, dual controls. Returning, the mechanic put the case of oil on the empty seat beside him. He was on final, approaching the Red Lion runway, when the plane hit a bump and the heavy package jumped off the seat and rammed the stick forward. The Cessna plunged to the ground, nose first. The mechanic was killed. Linkenhoker took the seat that the oil had been on and emplaced it in the cockpit of the 26. To this pattern of aircraft construction by junk collage there was one exception, the control system—rudders, elevons, cables, hardware—all of which was new. The engine, though, was the same four-cylinder, horizontally opposed, two-cycle McCullough that had powered the triple-hulled Aereon.

Linkenhoker used tools from his workbench at home. He cut metal with a wood saw. "It's pretty hard for people to believe—this back-yard project," he said at one point. "When you try to describe the shop that you've been working from and the tools that you've been using, people smile with a gentle little smile and turn away to keep from hurting your feelings. I've had my doubts. I've felt somewhat insecure. It doesn't worry me any. This is the type of work I like to do. It's in the airship field, which is the only thing that I really know. This is a stepping stone to future airships. Just how big a part it will play I haven't the slightest idea. I wouldn't even try to prophesy." He had been in naval aviation for twenty years, almost always in airships, sometimes flying as a crewman, doing structure and envelope rigging, on patrol from Maine to Brazil. He had gone into the Navy straight from high school in Covington, Virginia, where his father was supervisor of a machine room in a paper mill. The young sailor developed expansive feelings about the future possibilities of lighter-than-air, and with the others—the other helium heads—he gradually became disillusioned and, ultimately, bitter in the feeling that the Navy had sold them out. "Any form of aviation must be experimented with," he would say afterward. "L.T.A. experimentation was absolutely nothing. All through the Second World War they should have been experimenting, but they were flying the same type of airship up till the very last. This is a touchy subject. It was strictly a one-sided affair. We were down to twenty ships by 1953. Nothing new was cranked into the picture; it just had to reach its end." When the Navy gave up airships, Linkenhoker gave up the Navy. He had nowhere to go where he might have faintly cared to go except to the Goodyear Tire & Rubber Co., which kept a few blimps alive, like the whooping cranes in the New Orleans zoo. The Goodyear blimps carried fields of light bulbs through which rippled advertising messages to America in the night. Linkenhoker was ready to settle for that—to travel from city to city, away from his family, as airborne maintenance man in a hovering billboard—rather than lose touch forever with airships. He was virtually packed for Akron when Admiral Rosendahl called him and suggested he go to Aereon. It had taken him six years to build the triple hull, and three more to build the 26.

Charlie Mills pulled his head out of the hatch, and Linkenhoker went on with his inspection, experiencing a refreshed sensation of relief that Mills was no longer the test program's test pilot, as Mills had been once, at Red Lion. When the 26 had made its first taxi runs, two years earlier, it had no skin. Rolled out into the open, it was a jungle gym even beyond the imagination of children, and it moved up and down the Red Lion runway at forty-two miles per hour through fresh snow, with Mills, dressed in an astronaut's puffy flying suit, perched in the front, like a bird on a naked branch, his feet sticking forward into a cold wind. The ship without skin could not possibly leave the ground. What worried Linkenhoker and everyone else was what might happen when the aerobody was covered with cloth. It had become apparent that, given half a chance, Mills might shrug off the plotted prudence of the test program, eschew the developing counsel of engineers and computers, turn up the engine, take the stick in his hand, and—wings or no wings—bolt for the sky. "There are two ways to test," Mills had said. "One is slow, gingerly, step by step. That way consumes time and money. The other approach is short-time, high-risk. With the 26, the best procedure is to fly it. That's what pilots are for— to take risks." Linkenhoker covered the ship with airplane cloth, and Miller told him to paint it Princeton orange. Linkenhoker went to Shick Auto Supplies in Trenton, and said he wanted Princeton orange. "Mack Truck orange is what we've got here," the man said, and Linkenhoker said that would do. To save weight, he applied only one coat. He used no primer and only five coats of dope instead of the normal nine or ten. When Charlie Mills was about to enter the cockpit to perform another taxi test, Miller, visibly nervous, took him aside and said to him, "Charlie, you are not to fly the 26." Hundreds of items had to be proved out, in many taxi tests, before a first flight could even be contemplated. The replacement cost of the 26—time and materials only—was about a hundred and fifty thousand dollars.

"Wilco," said Charlie, and down the runway he went and took a little jump into the air. The wheels rose only an inch or two off the ground, but Miller aged ten years and Linkenhoker bit his toothpick in half.

Now, at NAFEC, standing among the expectants around the 26, Mills said, "Had I been running the railroad, it would probably be a wrecked aircraft by now." There were appreciative grins. No one spoke up to disagree. Linkenhoker, out of sight inside the ship, thought of the catastrophe that might have occurred because "Charlie wanted just to jump it into the air and try a go-round." Linkenhoker liked Mills well enough, and unreservedly admired his skill in airships, but he sensed the complexities within Mills—fast assessments, speed-drying interests—and, although he could never feel it was his place to say so, he preferred seeing other characteristics in the cockpit of the 26. Olcott was modern, circumspect—more mathematics, no flair. Olcott had already taxied the 26 more than a hundred miles. "I like the cut of his—his attitude," Linkenhoker would say. "His matter-of-fact reasoning. There never seems to be a need for words between Olcott and me. There's an understanding there that doesn't need words. I know what I have to do, and he knows what he has to do."

Olcott had changed his clothes. His jump suit was green. His flying boots, actually construction worker's boots, were new ones from Sears, Roebuck. He put on a white plastic helmet marked "U.S. Air Force." Briefing, he said to everyone present, "We're going to try a lift-off in ground effect, a short hop; and, if that works out, we'll try a lift-off and a prolonged straight and level flight, still in ground effect." The memory that the 7, repeatedly trying to do just that, had porpoised, oscillated, jackrabbited, and finished its days in smithereens had clearly been screwed into a remote corner of Olcott's mind. His computer—at his firm, Aeronautical Research Associates of Princeton—had helped him do this. The computer, full of equations of motion that mathematically detailed the performance characteristics of the 26, had been refreshed with the data gathered on the final flight of the 7. Olcott had then spent four days flying the computer, moving an actual control stick that changed the variables of the equations. Voltmeters plugged into the computer could be read as airspeed indicator, pitch-attitude indicator, rate-of-climb indicator, turn-and-bank indicator, altimeter, and so forth. Moving the stick, Olcott had tried one variable at a time—roll damping, roll rate, pitch damping, vertical acceleration, horizontal acceleration, control sensitivity. Basic truths of the deltoid Aereons were uncovered or confirmed. The initial response from up elevator would be a pitch change and not an altitude change, for example. While an ordinary airplane turns principally with ailerons, the primary control in the aerobody was apparently the rudder. If power was reduced, the nose would go up—also the reverse of what happens with a conventional plane. One way to get the 26 into a takeoff attitude might be to reduce power. In any case, the stick was not to be used to force the aircraft into the air.

Linkenhoker emerged from the 26 and said he thought everything was ready. Olcott said, finally, "I hope no one will be disappointed in what they see today. Nothing sensational will happen—I trust."

Holding a handle that was attached to the nosewheel gear, Linkenhoker began to tow the 26 like a huge wagon toward the hangar doors. A loud buzzer sounded as power came on and the doors began to part. Boeing 707s had often rolled through those doors with no clearance problem whatever. The 26, whose trailing-edge control surfaces had extended backward beyond original intentions, was twenty-seven feet seven inches long (Pitot boom excluded). Behind its pinpoint nose, it widened to a breadth of twenty-four feet across the tail. So the big doors opened scarcely a crack, and the 26 went out into the night.

Lights of the airfield, blue and red, spaced out through the dark to the horizon. Airplanes parked on the apron formed high silhouettes behind the 26, black on deep gray —a giant Globemaster, its engines hanging like teats; a Boeing 720, slick as the night; a couple of Convairs; a DC-7; an Aero Commander; a DC-3. A bar of light from the aperture in the hangar fell across the 26. Olcott climbed through the hatch and into the seat, his white helmet appearing to be almost phosphorescent within the plastic bubble. He strapped himself in. He strapped his note pad to his right thigh. Now that he was aboard, cast-iron weights were removed from the nose, and a Detecto bathroom scale was slipped under the nosewheel, read, and removed. Dawn began, pushing pink streaks upward, painting out the eastern stars. Venus stayed. Gray lightened and turned into high pale blue, with white bits of cirrus in it. The pinks flared. The dark airplanes on the apron turned silver and cold. White stucco buildings now stood out across the airfield flatness. One of these was a firehouse. Its doors opened and fire trucks began to move toward stations on the runway. Red as blood, they had enormous gold numerals on their sides, the figure 6, the figure 8. A red fire car, sedan of the chief, sprinted cross-country toward the 26, stopped nearby, and waited, its roof light revolving and flashing. Station wagons drew up carrying flagstaffs and large, flapping orange-and-red checked flags. One station wagon was Linkenhoker's Dodge, which he would drive. The other was the airport operations car of NAFEC—roof light turning, flashing, now red, now yellow. All vehicles were equipped to radio the NAFEC tower. Olcott lowered over his eyes the glare shield of his crash helmet. Putman leaned in through the hatch. Bits of conversation drifted out.

"Fifty-six knots calibrated is the best rate-of-climb speed," Olcott said.

"I concur."

"Alpha R is seven point six. Delta E flight is eight point four."

"I read it nearer ten point zero."

"O.K."

"You could have a three-point lift-off. Do you understand what I'm saying?"

"Yes."

Putman, aerodynamicist, who had never flown or cared to fly, was trim and handsome, with rich dark hair that was now tumbling across his forehead. He was wearing a blue shirt and a gray sweater that had a hole in one elbow. His trousers were vertical candy stripes—blue, white, and gold. A comb protruded from a hip pocket. Putman had built models all through his youth—Fort Smith, Arkansas (he was an undertaker's son), Phillips Andover Academy, Princeton University. He never tried to fly the models he built. What mainly attracted him about airplanes was their incredibly beautiful appearance, airplanes as pinnacles in the aesthetics of function. He was thirty-six now, advanced-degreed and academic: Princeton Department of Aerospace and Mechanical Sciences. His wife had once filled in a coupon that brought an Encyclopaedia Britannica salesman to their door. The salesman was Paul Shein, Aereon's treasurer. Putman bought the encyclopedia, and Shein drew Putman into Aereon as a consultant.

"I'm basing fifty-two indicated as fifty-six calibrated," Olcott said. "Fifty-two indicated gives me a reasonable takeoff solution."

Putman said, "Very well."

"The first run will be without rotation. The second run will include a rotation at a forty-seven-knot takeoff solution. The third run will be a go."

"Jack, you're wise to select to do the rotation at a lower speed," said Putman.

Olcott said, "The only thing I'm concerned about is staying on the front side of the power curve."

"O.K., then?"

"O.K. I'm all set."

Linkenhoker, at the rear, stood on a two-step steel ladder and put his hands on the propeller, which had a forty-eight-inch diameter. A self-starter would be dead weight in the air, an impossible luxury at this level of the aerobody's development. Linkenhoker shouted, "Contact!"

From the nose, Olcott shouted, "Contact!"

Linkenhoker pulled down hard on the propeller. It turned a quarter turn and nothing happened. He tried again. Nothing. Again. Nothing. Again. Nothing. He shouted, "Full throttle!"

"Full throttle!"

Four more tries. Nothing. Linkenhoker paused, and rubbed his head against his shoulder. He primed the carburetor. He tiied six more times. The engine did not so much as cough. Miller displayed signs of nervousness. Mills, who had a stopwatch in his hand, looked impatient. Putman folded his arms and waited. Linkenhoker seemed annoyed—nothing more. There was very little he did not know about motors.

"Do you think it will start, Link?"

"Of course it will start. It's a gasoline engine, isn't it?"

Reaching high, he jerked the prop down with all his weight and muscle. Nothing. In quick succession, he heaved at it five more times. Once, it eructed slightly; but that was all. Linkenhoker's hair was becoming damp with sweat. "Switch off!" he shouted.

"Switch off!"

"You want more priming gas, Link?"

"No, I want some more breath." Linkenhoker suddenly looked vulnerable, as if he, not Olcott, were the man in danger. He was near fifty and not, by appearance, in condition for this much of a workout. "Switch on!"

"Switch on!"

Five more tries. Nothing.

"Half throttle!"

"Half throttle!"

One more try. Nothing. A gill of primer. Another heave. The engine started. It sounded exactly like a chain saw. It sprayed noise off the high walls of the hangar, killing talk. Olcott closed the hatch and called the NAFEC tower. There was no response. His radio was inadequate. He opened the hatch. Could someone else please call the tower for him? Soon, in the NAFEC operations car, a thumb went up. Olcott raised a thumb in acknowledgment, and the 26 began to move out past the big silent jets. Linkenhoker drove Miller and Mills to the main runway, and they got out on the turf at the edge, midway between the ends—Mills, the consultant, with his stopwatch; Miller, the president, maintaining the corporate records with his Nikon Super 8. The station wagon sped away. Hundreds of seagulls were holding some sort of meeting on Taxiway Bravo, parallel to the big runway, and Linkenhoker's Dodge scattered them as he hurried to join the 26 and the other vehicles in time for the first run.

The dawn air was cold. Mills wore a suède jacket, old khaki trousers, old flying boots. Miller wore his Navy flight jacket, which had been issued to him twenty-five years before. His past seemed somewhat at odds with his theological present—the Master of Theology whose time-and-a-half labors for Aereon were virtually equalled by his continuing church work, as, for example, a leader of the

Children's Sand and Surf Mission, in Ship Bottom, New Jersey. Large block lettering on his flight jacket said "ATTACK 35," on an emblem in which a fire-breathing flying red dragon was riding a torpedo. Miller, however, had long since reconciled the divergences in his cosmography; witness a psalm he had once contributed, as a Naval Reserve fighter pilot, to the *Bulletin* of the Officers' Christian Union of the United States of America:

Savior, you have launched me, and you will bring me aboard,
    I shall never land short.
    I can never land short of your flight-deck;
    short of your elevators are the deep waters of death.
    The black waves you have stilled;
    you have overcome them that I may land in heaven.
I shall never land short, for you, my Savior, have launched me,
    and you will surely bring me in.

The sun, now a deep-red full circle lifting from the horizon, was halved by a wafer of cloud that appeared black against the red and against the pinks and blues of the sky. The stars were gone. The air was sharply clear. A six-knot breeze had been predicted for this hour, but even that had failed to develop. "We must thank the Lord," said Miller. "This day is providential. He has given us perfect conditions. Clear. Dry. Dead calm." Fearing observation, Miller scanned the peripheries of NAFEC—the chain-link fencing, the automobiles moving on exterior roads. The ones that moved slowly made him nervous.

The 26 crescendoed, wheels rolling firm on the ground, in a high-speed taxi run. In single file, the station wagons, the car of the fire chief, and a fire engine raced down the taxiway—flags flying, lights flashing and whirling—keeping pace with the 26. These five vehicles were the only moving things in the huge level acreage of NAFEC. At the far end, Olcott turned the 26 around. Having neither seen nor felt anything that he did not expect, he decided that that was enough taxiing and he was ready to fly. He went over his checklist. Both boost pumps on. Twenty-one minutes of fuel consumed. Controls O.K. Very simple. Forget nothing. Next, he reviewed his plan for rotation—the moment when an aircraft, on the ground and rolling, lifts its nosewheel and assumes an angle of attack from which, with added acceleration, it will rise into the air. He would rotate at fifty-two knots with an elevator deflection of thirty degrees. This should produce an angle of attack of eight degrees for initial lift-off. In flight, the angle of attack would increase to ten degrees. He would hold that attitude, flying the aircraft. He looked out through the clear-plastic canopy at the other vehicles all around him like sheepdogs, protection itself, watching, waiting for his move, within his philosophy of small increments, of small and careful steps one at a time, toward a new level. He looked at the NAFEC operations car and raised his right thumb. The tower was told. A thumb went up in the car. Olcott accelerated to full throttle.

On the windscreen before him were two horizontal strips of black plastic tape, one above the other, like an equals sign. The upper tape was on the line of sight between Olcott's eyes and the horizon. The other tape was ten degrees lower. Gaining speed, he scanned his indicators: airspeed, angle of attack, control position, revolutions per minute. At fifty-two knots he deflected the elevators and watched the horizon descend, steadily, fairly rapidly, just as it should do, from the one tape almost to the other. In rotation. Angle of attack: eight degrees. One hand on the throttle, the other on the stick, he held things just where they were, and he began to ask himself questions. How much stick force is necessary? Is there sway? Is it heavy on the main wheels? Is it spongy? He was feeling his way toward the first indication of lift-off. He thought of nothing else and noticed nothing else—not Mills beside the runway, smoking a cigarette, his stopwatch running; nor Miller, following the aerobody with the Super 8 like a duck hunter; nor the file of automobiles, bouncing and lurching at high speed over the ground swells of the taxiway, racing to stay with him, while engineers within them narrated the scene into spinning cassettes. He asked himself if the 26 was responding. Did it have a mind of its own? Did it tend to stay where he put it? He sought motion cues. Vibration. Acceleration. Runway shock. There was now no runway shock. Fifty-six knots, and the 26 was airborne.

Putman, riding in Linkenhoker's car, said, "I see daylight! I see daylight under the wheels! Estimated angle, ten degrees." Linkenhoker chewed his toothpick, drove the car, and said nothing. Miller, by the runway, shooting film steadily, did not dare to expect anything after eleven years. The moment brought to his mind an image, as he later put it, "of long bedraggled nights at Mercer County Airport, with dead flies on the floor, Linkenhoker working late, night after night—a whole fragile structure brought to focus: torn between ultimate trust that all things work together for good for those who are the called (my own overarching awareness of the providence of God) and authentic human stress." Mills grinned, standing there in his old flying boots, his reading glasses dangling from his neck. As the 26 came at him and went past him, he said, "Steady as a rock. I haven't seen an Aereon fly that steady since the old rubber-band model."

The 26, as the computer had said it would, moved its own angle of attack, after takeoff, from eight to ten degrees and firmly held it there. No wild oscillations for this ship. No need to manhandle it. Everything seemed to Olcott to be on nice, manageable terms. The 26 had slid into the air. Felt just like that. It eased into the air. In the right lateral-directional sense, it sat there. It was very soft. It felt very smooth. He reduced power. Kept everything else the same. Felt for the ground. Maintain the body attitude, he told himself. Feel for ground contact. He was concerned about ballooning into the air. He searched for the first indication that the wheels were on the ground. He asked himself, Is the pitch effect right? The runway gave him his answer. Main wheels down. Power change. Adjust for attitude change. Roll out. He came to the end of the runway. He turned the nosewheel. The 26 spun around, and sat still, pointing back into the airspace it had come out of, having its day as a falcon, spreading the wings it did not have.

It had flown a thousand feet at an altitude of twenty-four inches.

Mills said, almost to himself, "What a beautiful flight!" Watching it, he had multiplied the size, the altitude, and the range of what he had seen. He had seen a rigid airship, with new proportions, new missions, flying not a thousand feet but a thousand miles. The rest was detail. Work it out, men; on the double. Mills might have been an admiral if the lighter-than-air program had survived. One of his last acts in the Navy had been to write a thesis for the War College called "Airships—Renaissance or Requiem." Then, like many others, he faded into the sycamores of a small New Jersey town—a picket fence, a frame house, big airships visible, if nowhere else, in pictures on his study wall. He kept a bottle of Windex on his desk. He cleaned his glasses with Windex, the better to see the pictures of the blimps. "This was the M. The sweetest ship that ever flew. Look at that long car—a hundred and seventeen feet; in effect, a dorsal fin. She's more stable in yaw than any other airship I've ever flown. This is me at the controls. And here's an idea we had—a helicopter pad on top of an airship. We actually had it half built. If new airships were constructed today, we would have a line waiting to use them tomorrow, for radar calibrations, for atmospheric samplings, for ecological surveys, for airfreight. Airships used to use goldbeater's skin (ox intestines) for gas cells. Engines were built at eight pounds per horsepower. Now we could do three-quarters of a pound per horsepower. Think what could be done with modern materials. We came to an untimely end." He pressed the stopwatch. Olcott, a mile away from him, was starting another run.

Don't step out of the straight and narrow, Olcott told himself. Don't get carried away. False confidence can attend the second run, so this is where I have to be careful. Then he drove the 26 up an invisible ramp until it was ten feet off the ground, where it remained in level flight without a quibbling motion while he diverted his attention to collecting data. After flying two thousand feet, he landed, smooth as talc. Pausing, making notes, he almost quit for the day, because he wondered if he had become overconfident. Deciding that he had not, he took off again. This time, he allowed his left hand to leave the controls, and with it he clicked pictures. (A camera was mounted above his left shoulder and focussed on the instrument panel.) The flight was again perfect, nose high, ten feet up, three thousand feet from lift-off to touchdown. Mills clicked his watch, and said to Miller, "Don't stop now, Bill. Don't, for heaven's sake, stop now. Have the test go on. Conditions are right. Everything is right. Do the next phase now."

"I'll speak to Jack," Miller said, but there was really no need to. He could already hear Olcott saying, "I want to go into unknown areas on my terms, with no surprises." In fact, the 26 was already moving toward the hangar. Miller and Mills walked across the runway, and for what had happened Miller silently offered thanks to the Lord. The heel of one of Mills' boots came off and settled like a coin on the asphalt. He picked it up, and took off the boot, and walked with one shoe off and one shoe on, attempting repairs. "All my eight thousand hours of flight are in these shoes," he said. "All my eight thousand hours of flight." With his fist he punched the heel back on. "Don't stop now, Bill. Do the next phase now."

Miller was swelling with exuberance, but Calvin within him would not let it brim over. On the apron beside the hangar, Miller suggested that Olcott and Linkenhoker pose for photographs with the ship the one had built and the other had flown. Olcott looked uncomfortable. Linkenhoker—much too shy for show business—blushed, bent his head forward, took hold of the 26 by the nose, and dragged it into the hangar. The debriefing, held around a table in the NAFEC cafeteria, was in its way like the debriefing after the last, and most spectacular, crash of the 7. Everyone's voice was flat. The discussion was totally technical. Neither joy nor dismay had relevance there. Developed facts were what mattered. "The sequence was to accelerate on the ground with the controls faired," Olcott was saying. "The controls felt a little mushy, but they're responsive. I had no difficulty setting the angle of attack. The landing was accomplished by maintaining the same angle of attack and reducing the power. Letting the 26 settle at that attitude is just the way to do it."

"We need a ratio of predicted to actual values, and then we can decide how far to back off the tab," Putman said.

Olcott nodded in agreement. "I would say we're right where we ought to be," he said. "Tuesday morning, we'll explore prolonged hops in transition out of ground effect, and prolonged hops out of ground effect."

Afterward, Olcott gave me a lift back to Princeton in his Karmann Ghia. "We're exploring relatively unknown areas," he said at one point. "It's not an airship. It's not an airplane. It's what Bill Putman says it is: an original concept—something like the lifting bodies that NASA tested for reëntry vehicles to land on land. When we lifted off today, it was just a very, very small step into an unknown region, and we had been there before in the piloted analogue simulation. I felt no big heartbeats or adrenalin inputs. The first flight was a normal extension of what we had done beforehand. No quick judgments. Right on plan. I didn't want to dump on Miller's elation, but for me it was business as usual. No hurrahs. The thing that pleases me is that we were able to predict what would happen. If the predictions were correct this time, we can have more confidence in other predictions to come, when we try to take the vehicle a little farther." Olcott suddenly showed alarm, a trace of panic, the only trace of panic I was ever to see in him. His gasoline gauge was riding on "Empty," and had been for who knows how many miles. He drove on for ten minutes, a little tight in the lips. "I do this often," he said. "I just forget." An Amoco station saved him. Sitting there beside the pumps, he took out a notebook and recorded the current mileage and the re-placement volume of the gasoline.

C ELEBRATION was even less in order than Olcott might have thought. The 26 had got off the ground, but that was all it could do. In successive tests in following days, it flew on the ground cushion, never more than fifteen feet above the runway. It could not get out of ground effect. The increased drag in the zone of transition out of ground effect was too much for it. It was trapped like an electron in a magnetic bottle, a live mosquito under glass. The 26 would make straight, low, steady flights of about a mile, one after another, always settling back to the runway. Olcott's small steps forward into the unknown grew ever smaller, as the amounts of freshly acquired data diminished with each run. The test program went into a cycle of getting nowhere—the long rides in the dead of night to NAFEC, thermos coffee, the gradual breaking of the glaze that comes with getting up at 3 A.M., the brightly lighted hangar, the briefing, the roll-out into the dawn, moon hanging there as orange as the aerobody. The start-up. The coughing engine. The fire trucks, checkered flags, whirling lights. Flight. The 26 fighting against prohibitive physical odds. Scraping against an aerodynamic ceiling. Unable to climb. End of outing. Debrief. Go home. Nothing much accomplished. NAFEC people sometimes looked at each other in amusement after watching the so-called aerobody in action. On the same runway, they had seen McDonnell Phantoms, capable of going sixteen hundred miles an hour, and Lockheed U-2s that could climb thirteen miles into the air. Now here was something that took off like a snail going up a grape leaf and could not go higher than fifteen feet—an aircraft that could be outrun by a fire truck. The woman behind the counter in the NAFEC cafeteria, where windows gave a panoramic view of the runway, looked through tiers of glass shelves in front of her and remarked that the 26 regularly got up into the English muffins but seemed to get stuck there. "What do they think they're doing out there?" "I doubt if *they* know." "Beats the living hell out of me." "They call it an aerobody." "Milk or lemon?"

The test program was operating within margins that were narrowing. It had been agreed, for example, that the engine would not be used more than twenty-five hours. It was just a drone-aircraft engine, and an old one at that, but, within the concept of a slow-moving aircraft cheap to run, the engine's ninety-two horsepower was supposedly and necessarily sufficient for this stage in the evolution of the aerobodies, no less than the power in a rubber band had been for John Kukon's smallest of Aereons when it flew at Mercer County Airport. To replace the engine now, or even to overhaul it, would queer the mathematics and shatter the data already collected. The engine had been used twelve of the twenty-five hours.

Fatigue failures had to be pondered, too. The 26 had not been built to fly the Hump. It was the "proof-of-concept model" in the long chain of Aereons, and once it had tested the concept it would have done all it could. Miller had called upon a Long Island aeronautical consulting firm to do a stress analysis of the 26, and they had sent a young engineer named John Weber, who had a saturnine expression, piercing blue eyes, and an obvious love of what he was doing. "I had a dream," he said one morning. "I dream about this ship sometimes. I'm going to check out those antennae." His left arm was broken, from a motorcycling accident, and he wore an Ace bandage around it while he worked, with Linkenhoker, on the 26. "There are fatigue failures in some of the weld surfaces, caused by droop loads," he told Olcott. "All your failures so far have been fatigue-type failures due to vibrations and other motions in the various systems."

"So you're suggesting keep the running time down."

"Exactly."

The effect of all this on Miller was to heighten his nervousness—deadlines rising over him like halberds. Surrounded by friendly mercenaries, haunted by imagined adversaries, he saw that it was up to him to do the worrying, for he had Linkenhoker on salary and everyone else was a paid consultant. Miller, by now, had a third of a million of his own dollars in the project and the tenth part of his life, and while consultants directed his show he could only stand by with a Super 8. When they talked strobes, tachometers, L delta A, he could comprehend, more or less, but he could not contribute. They had facts. He had faith. They consulted computers. He talked to God. They got direct answers. He did not. His reaction was to generate more faith. He would soon have to generate more money as well. He fretted. He lost too much sleep. His eyes were rimmed in red. One of the vertical fins seemed to him to have suddenly developed an alarming concavity. He caressed it with his hand. "I notice a flat spot here," he said. "Was it always here?" Weber looked up and nodded, speaking across the memory of dozens of such questions. "Yes," he said. "The flat spot was always there." Miller was too polite to become a rampant nuisance. He retreated, for the most part, into his nerves. His fears ran to a deep source and often brimmed over. He spoke sotto voce in restaurants and did much peering around. He could be having eggs in a diner and imagine that the truck driver sitting on the stool next to him was straight out of the spy bin at Boeing. He suspected anyone in the NAFEC cafeteria who happened to sit down at the next table. "Careful," he would say. "We have a listener." He wrote security memos and distributed them to the group: "We continue to require privacy and your help to achieve it. 'No comment' will be your response to inquiries. . . . No photographs are permitted. . . . No visitors are welcome except as invited by Management." He mimeographed a non-disclosure agreement, a promissory note of eternal mumness, and placed it, asking for a signature, before almost anyone who happened to glance at the 26. "Miller is psycho as far as security is concerned," said Mills. "He makes you sign agreements that you will not smoke cigars shaped like airships." Miller had his reasons. "I want to preserve our lead time" was the way he put it. Big companies would discover the 26 soon enough. Moreover, he felt that Monroe Drew, Aereon's founder, had seriously eroded Aereon's credibility by not delivering on his ballyhoo, a mistake the company might not survive twice. Whatever the cause, he was so fearful that he always seemed to be looking under closed doors for the telltale foot of the invader, and out on the airfield at NAFEC he would scan the middle distance with slow and careful eyes. One day, Olcott, who had sharp eyes, too, opened the hatch after a run and told Miller that a fireman in one of the NAFEC fire trucks was busy snapping photographs. Miller was galvanized—fears confirmed. He reported the discovery to NAFEC. The film went out of the fireman's camera and into Miller's pocket. The fireman had wanted to show the flying pumpkin seed, weirdest thing he had ever seen on the job, in Kodacolor, to his children.

A propeller of a different pitch was tried. It was made of lemonwood. Pusher props are not common, and this one was virtually unique. Igor Bensen had used it trying for a world speed record in autogiros, but that did not work out, and Linkenhoker had found the prop "just lying around," as he put it, at Bensen Aircraft, in Raleigh, North Carolina. A second battery was added, because Olcott said an up-and-away flight would not be possible without it, and it was installed just enough to the rear to shift the center of gravity in that direction five-tenths of one per cent. With the Detecto scale under the nosewheel, Olcott and the fuselage around him now weighed five pounds less, the better to nose up out of ground effect and into the sky. Everyone was full of anticipation for the next takeoff.

These minuscule changes could make the difference. If derision was rising among the personnel of NAFEC, it was possible that they were too used to watching experimental aircraft that were pushing at the upper limits of aerodynamic possibility. This one was at the inverse frontier. Its ambition was to become the Super-Slow Transport. It wanted to burn next to no fuel, in low-cost, low-power engines. Its immediate goal was not merely to fly but to fly within severe conceptual limitations. Future aerobodies would be large enough to take advantage of lifting gas, and minutiae would matter less. Meanwhile, the economics underlying the whole Aereon idea could be assayed in the 26, and fine points were extremely critical: the pitch of the prop, the weight of a few gallons of fuel, the pilot's choice of flying clothes, a little extra fat in his dinner.

Now Olcott moved the engine up to thirty-eight hundred and fifty revolutions per minute and began his bid to get out of ground effect. Linkenhoker, toothpick dancing, raced along the taxiway in his Dodge full of engineers. The taxiway was uneven, many dips and rises, and suddenly a congress of waddling seagulls was directly in front of the car. "Man, those birds are slow," Putman said as the car sped into them. Weber said, "It's early in the morning. They're tired." Somehow, Linkenhoker missed all but one, which jumped into a headlight and caromed off in a spray of feathers. A man in the fire truck, close behind, laughed uncontrollably. The 26 was airborne. Everybody in the Dodge sat forward expectantly and looked for sky beneath the landing gear. It never came. "He's not at altitude. He's not at altitude," Weber said. "I don't think he knows his altitude." The 26 was flying, as it would fly on to the end of its traditional mile, six feet off the ground.

Olcott opened the hatch. "There just isn't any more up stick!" he shouted. There was no point in trying anymore. As he taxied to the hangar, the station wagons were lined up funereally behind him.

Linkenhoker said, "I feel like five pounds of manure in a two-pound bag, that's what I feel like." He drove past the waddling gulls, and skirted them this time, and said, "I think we've got a very sick bird out here somewhere."

In the cafeteria, Olcott ordered doughnuts and smiled gamely when he was reminded that he was still flying in the English muffins. The group—Weber in a T-shirt, Miller in his flight jacket, others in sports shirts, Olcott in a charcoal-gray suit—sat at white Formica tables and reviewed their problem. Outside the windows, ponderous Starlifters from McGuire Air Force Base—seventy-ton aircraft, wings drooping—made touch-and-go runs where the 26 had been struggling half an hour before in its attempt to make the Starlifters obsolete.

"We need at least forty-one hundred r.p.m., and we're just not getting it," Olcott said.

"It sounds as if it's peaking out," said Weber. "We can strobe that prop. The tachometer might be off."

Linkenhoker said, "There is definitely a malfunctioning in that tachometer."

Olcott bit into a doughnut. He spilled some confectioners' sugar on his blue fleurs-de-lis tie. He brushed the sugar away. "Fix tachometer," he said. He wrote, as he spoke, in a neat hand on lined paper—a list of things to do before there could be another outing. He looked up.

"The engine is not putting out," he said. "There is something wrong with it, or the prop is at the wrong pitch. We need an engine fix or a control fix before we go out again. We need to explore a farther-aft c.g. It could be that this prop is just a little bit too high-pitched. The previous prop was underpitched. We may need a prop in the middle somewhere to give a little more power to get through ground effect."

"There is no strobe here at NAFEC," Miller said.

Weber looked at him in disbelief. "They don't have a strobe in a place like *this*?"

"O.K. Borrow one from Princeton, bring it down here, strobe the prop, and calibrate the tachometer," Olcott said, with discernible impatience. "We've got to move forward. A couple more outings like today's and the word is going to get around: 'Hey, they can't get out of ground effect.' "

Miller looked uneasily around the cafeteria. Charlie Mills, who was picking the right moment, made a characteristic suggestion that one more try be made, going for broke. He pointed out that the 26 was behaving like any overloaded aircraft, that a plane operating above its maximum gross would typically lift off in the way the 26 had and then fail to get out of ground effect. The 26 had a backpack parachute in it, fitted to the seat. How much did the parachute weigh? Twenty pounds. Cigarette smoke was billowing out of Mills' mouth. What the hell good was a parachute six feet off the ground? Olcott was more than willing to do without it. Take half the fuel out, Mills further suggested. Thirty pounds. The radio? Seven pounds. Hydraulic lines? One pound. The fire extinguisher? A pound and a half. The group agreed to throw out a hundred pounds in all—just to try to prove that the 26 could get out of ground effect. Let other problems take care of themselves later on. The debriefing ended. Miller, shoving his chair back, said, "This is very reminiscent of Aereon 7, this whole experience."

It was September 28th, two weeks later, before all the items on Olcott's list had been checked off and the weather forecast seemed good enough for the seventh test outing of the 26. It was called for dawn, September 29th. I rode down to NAFEC with Miller, and poured out coffee by the map light in his Mercedes. We looked up and saw leaden clouds forming islands in the stars. We thought the weather forecast might have been wrong. Miller drove alarmingly—a little more so than usual, making decisions somewhat more slowly than the speed seemed to require. He was in a brooding mood. "No amount of hope will fly a stone," he said. "I could not have lasted through this ordeal if I were not a Christian."

The sky above NAFEC was clear by dawn. The clouds had all moved over to one side and were trying to stuff themselves into the horizon. The sun lifted behind them. People deployed. The crash equipment ran around as usual, costing Aereon twenty-three dollars and fifty cents an hour. The air was cool, compact, almost cold. A light, steady wind was blowing from the north. Olcott made two or three low flights that were disappointingly like the earlier ones, but he was just loosening up, getting the vehicle around him again, getting his memories into line. Then he made his move, for something or nothing. The 26 had gone only nine hundred and twenty feet down the runway when, with a slight wobble, a wheel at a time, it jumped into the air. It was doing fifty knots. It climbed. Its wheels, seen from Linkenhoker's car, which was running along the taxiway, for the first time ever cleared the pines on the horizon and were in silhouette against the sky. The 26 climbed more—completely out of ground effect—and Olcott established a steady state of flight. He was flying at fifty-nine knots—fifty feet in the air. Linkenhoker's almost rigid taciturnity broke down. He slapped Weber on the arm. "How does *that* grab you?" he said.

Putman said, "If he goes any higher, we'll have to pressurize him."

Olcott began to indulge in considerable aileron activity—roll excursions, roll excitation. "He's getting the feel of it," Weber said. "I think he may have been a little nervous, but he seems completely at ease with the airplane."

"The aerobody."

"Yeah."

Olcott reduced power, and the 26 found its way down to the runway.

Putman said, "If someone were to ask me, I'd say that was the first flight of the Aereon."

Charlie Mills was all smoke and smiles. He reminded everyone, "As I said before—out the parachute, out the radio, change the c.g., and the mother will go."

The sum of what had been learned was that the 26 would fly when out of ground effect. The debriefing might have been called "the deflating." The ship had gone fifty feet into the air but was so underpowered that it was not going anywhere else. "That prop is a club," Putman said. No wonder Bensen set no fires with it. A new prop would have to be carved. Where would you turn, in the present age of aviation, if you wanted a custom-designed, custom-carved wooden propeller? Well, there was a man in Texas, another near Denver; but in all likelihood, in such a situation, you would talk to Henry Rose, in Lititz, Pennsylvania. Had anyone talked to Henry Rose? Yes. A new prop would cost five hundred dollars, would take at least four weeks in the making, and should increase the aircraft's power by thirty-five per cent.

Hundreds of bits of black woollen yarn, each four inches long, Scotch-taped to the 26 by Linkenhoker, had shown the presence of another problem. Air moving over and under the aerobody was not buttoning itself up properly when it came together again. Increased drag was the result. Vortex generators—small metal plates poking into the airstream—would correct the difficulty. They would have to be made, and welded to the fuselage. The list grew.

Miller listened in mounting gloom. He knew the cost of a month or two of down time. He knew that the 26 was flying headlong into the red, and, on principle, he would let the program move forward only on invested money, not on borrowed money. Months might well pass (nearly six months did pass) before the 26 flew again.

"The rudders are nice," Olcott said, concluding the debriefing. "The roll damping is weak, but right now there is nothing in the stability and control points that would prevent us from making a circuit of the field." A Starlifter touched down on the runway, making puffs of rubber-smoke, and picked up again into the air. "I'd say today was a very successful outing," Olcott said quietly. "As someone in here said, we got out of the English muffins and into the Danish pastry."

THE FRONT PAGE of the New York *Herald* of September 8, 1863, was entirely given over to classified ads and a large map of Chattanooga depicting the military situation there.

WANTED IMMEDIATELY—A FEW FIREMEN, coal passers, landsmen and green hands; also two colored men for a steamer going on the blockade; ship for one year $12.50 a month and $54 advanced. F. Gallagher, 174 South Street.

NOTICE TO DRAFTED MEN! Any man drafted into the service of the United States, desirous of obtaining a substitute, can do so upon very moderate terms by applying to Henry Lindensruth, 108 Greenwich Street, N.Y.

A RESPECTABLE YOUNG GIRL wishes a situation as a stewardess in a steamer to California. Call at her residence, 144 West 32d st.

Page 2 was almost all Chattanooga: "The Rebel Army Probably Outflanked—Bragg's Old Army Deteriorating—Ten Thousand Deserters Reported Within Our Lines." On page 3, things began to happen all over.

CHARLESTON—The Siege—Engagement Between the Iron Clads and Forts Wagner and Moultrie . . .

FEARFUL MASSACRE BY INDIANS—A special dispatch from St. Paul, Minnesota, says news has been received that a flatboat coming down the Missouri River was attacked by Indians, and all on board, twenty-five in number, were killed. . . .

IMPORTANT FROM THE SOUTHWEST—Generals Grant and Thomas have gone to New Orleans. General Grant will command all the Mississippi region. . . .

IMPORTANT FROM JAPAN—Anglo-American War on the Japanese—The United States Gunboat Wyoming Destroys the Japanese Steamer Sarsfield and Silences the Forts at Kanagawa. . . .

MEADE'S ARMY—A grand review of the Third Corps by General Meade takes place today. General Sickles is expected to be present. . . .

AERIAL NAVIGATION—We have this week the pleasure to record the success of the most extraordinary invention of the age, if not the most so of any the world ever saw—at least the greatest stride in invention ever made by a single individual. . . .

The mayor of Perth Amboy had built and flown a dirigible airship. With *Herald* reporters present, Solomon Andrews had flown his airship up and away from Perth Amboy common, and had "demonstrated to an admiring crowd the possibility of going against the wind and of guiding her in any and every direction."

This was twenty-one years before Renard and Krebs flew their electric airship La France, thirty-eight years before Alberto Santos-Dumont flew his No. 6 around the Eiffel Tower, and forty years before the brothers Lebaudy made their twenty-eight closed-circuit trips in Lebaudy I —milestones in the acknowledged beginnings of dirigible lighter-than-air flight. Andrews' airship, which was powered by gravity, consisted primarily of three cylindroid hulls, each eighty feet long, sewn together at their longitudinal equators and covered with varnished linen. The inventor, as pilot, stood in a gondola that hung on cords sixteen feet below the triple hull. From the federal government he had received a patent, and he had also been given a charter to establish the world's first airline—the Aerial Navigation Company, New York to Philadelphia and back. When he made his first airship, he decided that it represented the beginning of a new age of man, so, inventing a word, he named it Aereon.

Andrews was tall, with flowing blond hair, a Grecian profile. His stance was nonchalant. The Phi Beta Kappa key tapped against a flat stomach. He was a medical doctor, trained at the College of Physicians and Surgeons, which was then on Barclay Street, in lower Manhattan. He had gone to Yale. His father, Joseph Andrews, was the minister of the First Presbyterian Church of Perth Amboy. One Sunday when Solomon Andrews was seventeen, he looked out a church window during his father's sermon and became absorbed by the flight of a bald eagle, which was moving through the air without stirring its wings. He felt, he said later, that the simple secret of flight had been shown to him at that moment, and he resolved to place among his numerous ambitions in life the construction of a device that would imitate the eagle. His ambitions and interests were, as they developed, gallimaufric. He invented, among other things, the combination lock. To advertise it, he locked up a thousand dollars in a small trunk, took it to Wall Street, and offered to give the money to anyone who could pick the lock. The offer lasted two months, and no one picked the lock. He invented the wickless oil lamp. He invented a kitchen range for anthracite. As mayor, and also president of the board of health, he designed and built the Perth Amboy sewer. In barracks constructed by the English Army in the eighteenth century he established workrooms for the manufacture of his inventions, which also included a fumigator, a forging press, a velocipede, a machine to crack nuts. He developed an automated barrelmaker that could turn out five hundred kegs a day. For thirty years, under government contract, he made the locks—key locks—for United States mailbags. The fortune he assembled from all these inventions he poured into the development of his airship. The fortune, though, was not enough. He hunted constantly for investors.

After observing first hand the awkward attempt of the Union Army to use balloons for aerial reconnaissance, Andrews wrote to President Lincoln, asked for support in the construction of Aereons, pledged fifty thousand dollars' worth of his own real estate to show that he was in earnest, and guaranteed "to sail five or ten miles into Secessia and back, or no pay." The White House referred the letter to the War Department, which referred it to a pigeonhole. Andrews persisted. Eyewitnesses of the early flights wrote to Washington. Eventually, a commission was set up to look into the Aereon's potentialities. Andrews went to Washington with hydrogen-filled India-rubber models, which flew around a room and returned to his hand. The commission recommended "a suitable appropriation" for the development of Aereons; but then the war ended, the President was assassinated, and the government forgot Andrews and his airship.

Andrews found enough investors to sustain him at least for a time, and he went ahead on his own. Reviewing and refining his theory of dirigible flight, he decided that the triple hull was not the optimum configuration after all. He would prefer, now, a tremendous lemon seed—a cylindroid Aereon, somewhat stubby, pointed at either end. Probably because he wanted to attract more attention than he had attracted before, he built this second airship in a vacant lot on the southeast corner of Greene and Houston Streets, in New York. To explain how his Aereons flew, he published a pamphlet called *The Art of Flying*, subtitled *Without Eccentricity There Is No Progression*. The nucleus of his conception was simple. A plank rising through water goes sideways, following the line of least resistance. A balloon with an elongated axis would also move sideways, like the plank, following the line of least resistance. It would move rapidly—it would *have* to move rapidly—because gravitation ignores lateral motion. For example, if a spherical balloon and an elongated balloon of the same displacement were released simultaneously, each would attain an altitude of, say, a thousand feet at the same moment. The spherical balloon, though, would rise slowly and vertically, while the elongated balloon would shoot off sideways at a handsome clip, up an inclined plane. Andrews had found that speed was related to displacement. He could achieve lateral motion of about one mile per hour for every pound that his ship was lighter than air. Thirty pounds: thirty miles per hour. Throw ten pounds overboard at the start: fly off at ten miles per hour. At a selected altitude, valve gas—that is, get rid of gas (he used hydrogen)—and descend, heavier than air, down an inclined plane. Andrews tilted the airship's nose upward by stepping to the rear of the gondola and downward by stepping toward the front. His angle-of-attack indicator consisted of three marbles in calibrated wooden grooves. When the first marble moved in its groove, the airship was inclined five degrees. When the second marble moved, the ship was inclined ten degrees. Third marble, fifteen degrees. He steered with a triangular rudder that was covered with cambric muslin. To rise again, throw off more ballast (he used sand). Valve gas to descend. Rise, descend, rise, descend—always in broad synclinal parabolas. When hydrogen and ballast run low, stop at a depot for more gas, more ballast. Dirigible flight, no engine—that was the art of flying.

The New York *World* was impressed. It reviewed Andrews' flying career under a deep stack of headlines on June 24, 1865:

AERIAL

A PLAN FOR UTILIZING
THE ATMOSPHERIC OCEAN
THE AIR MADE NAVIGABLE
A SUCCESSFUL VOYAGE
IN THE BLUE EMPYREAN
MARVELOUS PERFORMANCES
AT PERTH AMBOY
A FLYING JERSEYMAN
ABOVE THE CLOUDS
BOREAS DEFIED
BY THE AEREON
&C &C &C

"There is no immortality for Montgolfier, Godard, or Nadar," the article began. "A gentleman from New Jersey has mastered the theory of interplanetary navigation. . . . By his aid we shall be able to bridge the rainbows and go picnicking at the height of Mont Blanc. His invention is the culminating endeavor of the history of aerostation."

Andrews had said to the *World* reporter, "When the whole thing becomes known to the public, I expect to lose credit for ingenuity because of its simplicity. But I mean to entitle myself to credit for faith and perseverance. When Kepler announced his great laws of the planetary system, he said, 'I can well wait a century for a reader, since God has waited six thousand years for an observer.' So may I well afford to wait God's time, as I have done, for the development of aerial navigation, since He has honored me as its inventor or discoverer."

The *World* concluded, "Very soon, if his hopes are realized, we shall have naval battles in the air, assignations in the clouds, evening sails above the foliage of the park, and airline railways with stations in the tree tops. When the Broadway railroad will be no necessity, and cigar smoke will get into nobody's eyes but the skylark's, and the overland route to California will be free from the reach of the Moyane, the Camanche, or the Pottawottamie."

Andrews was a long time getting his second Aereon into the air, but when he did, on Friday, May 25, 1866, he took three other men with him: Waldo Hill, a director of the Aerial Navigation Company; C. M. Plumb, corporate secretary; and George Trow, publisher of *The Art of Flying* and of *The Aereon, or Flying-Ship*, and vice-president of the company. Emmett Densmore, treasurer of the Aerial Navigation Company, stayed on the ground. The new Aereon rose from the vacant lot and nearly ran into the building across the street. It moved northwest, with the wind, to Union Square. There it turned.

"The moment was critical," the *World* said later. "The verdict of years of toil, thought, suspense, of a life-felt, life-wrought purpose was committed to the result. Changing her course, the gallant vessel, freighted with so many hopes, veered around as directed and bore on her unswaying, undeviating way, with tremendous velocity, annihilating space, and spurning the wind across whose path she rode, and whose advancing hosts she met and conquered. . . . She headed in a southwesterly line, along which she shot at a rate of less than three minutes to the mile. The wind, blowing quite freshly, came almost directly athwart the faces of the voyagers, and pieces of paper cast on the bosom of the air were wafted in a course contrary to that pursued by the machine, thus conclusively proving that, unlike balloons, the Aereon can proceed, if need be, against, and not slavishly with, the wind. . . . Navigation of the air was a fixed fact."

Picking up a by now familiar theme, the *World* exulted elsewhere, "Lovers can henceforth soar." The flight of the Aereon had been "a voyage of discovery through the azure ether that hung as a veil over the busy city. The ship easily and gracefully ascended to a height of some two thousand feet, or considerably more than six times the height of Trinity steeple, where, in mid air, the ceaseless hum of the city ascended with diminished and softened effect, and from whence the multitudinous mass of humanity that darken the streets appeared as Lilliputs, surrounded by houses of equal miniature extent, each intent upon his petty task or pleasure."

Andrews made another flight, on June 5th. This time, only Plumb went with him. The *Tribune*'s aerial-navigation man, whose report appeared the next day, was the emotional opposite of his counterpart on the *World*. "It seemed clearly demonstrated that the balloon possessed motive power of its own," said the *Trib*, "but it was at the same time apparent that the proper mechanism for using the power to its best effect has not yet been obtained. The voyagers continued to progress toward the north and disappeared in a cloud. We have no intelligence of their whereabouts up to the time of going to press."

A Healing Art—A History of the Medical Society of New Jersey gives as record that Dr. Andrews created "the world's first dirigible." Aeronautical historians have been a little more hesitant in their diagnoses, ignoring him almost completely, and wondering, no doubt, just how stiff a wind the Aereons could buck. That latter flight went up Long Island Sound, for example, and terminated in Brookville. The previous one ended in Ravenswood, Queens. Andrews had shown, though, that he could turn to the left, turn to the right, fly a closed course. His aircraft was, in a word, dirigible. A depression rolled over the economy just then, and, being out of money, Andrews had to give up his experiments. A few years later, he died.

THE AGE OF ASSIGNATIONS in the sky, picnics over the foliage, may not have arrived as quickly as reporters once hoped, but the age did come, and by 1936 regularly scheduled voyages by commercial airship had been going on for almost thirty years. "Voyages" was the word used on the tickets, even for short runs within a country, and the connotation of ships floating in the atmosphere was not hyperbole but modest truth. The *Luftschiffe* (nearly all of them were German) were longer than most ocean liners. In the history of air travel, their voyages were without precedent and, as time would tend to show, without sequel. They flew at about seventy-five miles per hour —generally at low altitudes, where they functioned best, in the heavier air—and they crossed the Atlantic in two and a half days. They were steady, and extraordinarily free of vibration. There were no bumps. A milk bottle turned upside down and set on a table in Germany was once carried across the ocean that way, and it never fell. Trips were quiet, almost to the point of silence—engines far astern. In high summer, windows were open all the way. People looked through them at night down long corridors of light on the sea.

June 23, 1936, around 10:30 P.M., the Deutsche Zeppelin-Reederei's LZ-129—the Hindenburg—left Lakehurst for Frankfurt am Main. Eight hundred and four feet long, silver, cylindroid, with a fineness ratio (length over diameter) of six to one, the Hindenburg was the climax and meridian of the big rigid airships. Inside his cloth-covered aluminum frame (in German the gender of airships was masculine) were sixteen vast balloons—rubber bags filled with hydrogen—and they lifted him slowly into the night. He turned to the northeast, crossed pinewoods and horse pastures, thickening suburbs, Raritan and New York Bays. Then he moved in low over Manhattan, and his course was framed by the boulevard lights of Park Avenue, running on and on to the top of the city. The Queen Mary, in the Hudson, strung with lights, grunted cavernously—once, twice. Other ships cut in with horns, whistles, tugboat blasts of jazz-marine. By now, the passengers in the Hindenburg were themselves lighter than air. Most of them were leaning out the windows. There were fifty-seven of them in all, including Rear Admiral Greenslade, Lord Donegall, Captain Schulz-Heyn, Adriel Bird, Mrs. E. M. Latin, Erich Warburg, Olga Boesch. Max Schmeling was there, with a black eye. Max had just defeated Joe Louis. Jean Labatut, a Princeton professor of architecture, was aboard. And so was Charles Dollfus, the French aeronaut, the free-balloonist. Labatut, all but hanging by his feet, pointed a 16-mm. movie camera down at the city and used the airship as a dolly. The film that eventually resulted was, in effect, surreal. Points of light were all that emerged, against a background of absolute blackness. The city lights were less concentrated then, and Labatut's film of New York, which he still has, is like a slow float through beaded stars and inhabited planets. Labatut finally went to bed, nodding *bonsoir* to his countryman Dollfus, who had been assigned to the same stateroom. Labatut had expected the cabin—in fact, the whole ship—to be full of handrails to clutch during aerial pitch and roll. There were none anywhere. They were unnecessary. The Professor fell sound asleep congratulating himself on his choice of transportation.

A month or so before, Labatut, on sheer impulse, had walked into the travel department of the Princeton Bank & Trust Co. and asked them to get in touch with the German Zeppelin Transport Company and seek passage for him on the Hindenburg. Each summer, he taught at Fontainebleau. Why not go to Europe this time in—as he put it—a flying sweet potato? He was sure there would be no berth left for him, but some days later the bank called him in and gave him what appeared to be a proclamatory leaflet, ten inches high. It was his ticket: *"Die Deutsche Zeppelin-Reederei G.m.b.H. übernimmt auf Grund ihrer Beförderungsbedingungen die Beförderung von Prof. Jean Labatut mit dem Luftschiff Hindenburg von Lakehurst nach Frankfurt. Fahrpreis—$400.00, plus $5.00 U.S. Tax."* The company had asked if Professor Labatut would like to inspect their airship beforehand. Yes, he would. He drove to Lakehurst—this was just before the Hindenburg's departure on an earlier crossing—and he met Dr. Hugo Eckener himself, the commander, and stood stunned at the sight of the airship, a hill of silver, enormous beyond imagining. Out of the sky came a German aviator in a small airplane. He taxied around like a housefly. He had been giving aerial-acrobatic shows in various places in the United States, and he wondered if he could have a lift home. "Why not?" said Eckener. A large hole was cut in the side of the Hindenburg. The airplane was put inside. The cloth was sutured. Departure was not much later than planned.

Labatut got up early in the morning on June 24th. The Hindenburg was now sliding over Penobscot Bay. Labatut filmed that, too. He was a rarefied sort of architect, who was to become a teacher of teachers of architecture, but he had the child's eye of wonder, and he was stirred not only by the grace of the big airship but, even more, by the sedate revelations it presented as it flew. At such low altitude, a detailed frieze was constantly evolving beneath him, and he filmed it—animals, houses, towns, forests—in scenes dominated always by the two signatures of the Hindenburg, its reflection on water and its shadow on the land. Sometimes the reflection appeared black on white water. When the reflection ran into a shoreline, the shoreline cleanly ate it up. The Hindenburg just telescoped and disappeared. The shadow—great, unimpeded beluga—was somewhere else. Labatut would pan his camera and find it, rippling over the Maine islands or across great tear-shaped log booms on New Brunswick bays or across a field of high northern daisies on Prince Edward Island where a herd of cows ran in terror before the encroaching airship. Max Schmeling leaned out a window, a still camera pressed against his good eye, and recorded these scenes, too. Labatut turned and filmed Schmeling. The airship moved across Newfoundland's trackless interior, bridging wilderness rivers and climbing high mountains. The Professor filmed the shadow of the Hindenburg on the Annie-opsquotch Mountains. After three hundred miles of Newfoundland forest, the ship slowly crossed over blue-white icebergs in Notre Dame Bay and moved on over offshore islands, and then left North America for the open sea.

The passenger decks were not contained in any sort of external gondola but were entirely within the Hindenburg. Windows were slanted outward. They were part of the configuration of the hull. And as if that did not offer a sufficient view of earth and sea, the company had placed two rows of non-opening windows in the belly of the ship, so people could look straight down. When the customers showed signs of vertigo, acrophobia, or common fear, crew members jumped on the windows to demonstrate their tensile strength. Sleeping cabins were inside, amidships, flanked, on the main deck, by the lounge and writing room on one side and the dining room on the other. There was an aluminum grand piano in the lounge, and a bronze bust of Marshal Hindenburg. Labatut thought the bronze was striking, because it looked so heavy. He said that if Hindenburg's head had been aluminum it would have seemed lighter than air. The smoking room, a deck below, was double-doored and pressurized, to repel any loose hydrogen that might inadvertently wander into the presence of flame. The cooking apparatus, which was electric, was similarly protected. Matches and cigarette lighters were

confiscated at the outset of each voyage. The Hindenburg
had been designed for helium, which does not burn and is
a natural gas found in significant quantity only in Texas
and Kansas. Helium, twice as dense as hydrogen, can lift
only about ninety-three per cent of what hydrogen can
lift—a factor that had to be taken into account in the engineering of the airframe. Helium for the Hindenburg
once sat in steel bottles on New York piers, but it was
never shipped. Rigid airships had been effective weapons
in the First World War, and Harold Ickes, Secretary of the
Interior, refused to let a cubic centimetre of helium go in
1936 to Nazi Germany. The Hindenburg's passengers,
floating beneath seven million cubic feet of hydrogen, were
all but unmindful of it, though, as they banqueted on
Rhine salmon, *Kalbshachsen*, lobster, and caviar under the
beneficent eye of Eckener, who presided at meals. Mrs.
Latin asked Eckener, "What do you do when you go into
a storm?" Eckener turned to her heavily. He was a thickset,
jowly man with short, bristly hair and piercing eyes—a
composite king of Prussia. His glance pierced Mrs. Latin.
He said, "Madam, I am not stupid enough to go into a
storm." Dollfus, the aeronaut, enjoyed special privileges
from his friend Eckener, who gave him the run of the ship,
bow to stern, and Dollfus took Labatut with him onto the
catwalks among the gas cells. "I pressed my finger into the
rubber balloons and dared the hydrogen to explode," Labatut remembered, years later. "I knew what the hydrogen
was, but I was a true—a true innocent. On the catwalk was
just as if you were in a barn." Eckener invited Dollfus and
Labatut to the control gondola, the *Führergondel*, which
protruded below the keel. Two young crewmen were in

there with Eckener, each manning a ship's wheel. One wheel dealt with pitch, the other with roll, and it was up to these two youths to keep the Hindenburg stable in the air. Both were shining with sweat as they furiously whipped the wheels one way, then the other, panting, grunting, giving all they had to satisfy the gauges in front of them and the commander who stood watching. "That is *good* for them," Eckener told the two Frenchmen. "Very *good* for them." He also said that the duty at the wheels was so demanding that the helmsmen had to be replaced every thirty minutes.

Eckener inspired confidence. He was not just a flawless skipper. He was the leader of the airship transportation company, and—the ultimate credential—he had been trained by Count Zeppelin. Ferdinand von Zeppelin had conceived and developed the rigid airships. With the turn of the twentieth century, airship development splayed out in three directions. The sausage-shaped dirigible balloons of Alberto Santos-Dumont evolved into blimps, or non-rigid airships. The Lebaudy airships—long rubber bags with attached gondolas—evolved into semi-rigid airships (blimps with keels). Meanwhile, Count Zeppelin, of Friedrichshafen, decided that he would like to construct airships far too large to retain their shape on pressure alone. He built a huge floating hangar on Lake Constance, near his family estate. He hired metalsmiths and set them to work making transverse frames that would be joined together by longitudinal girders and covered with cloth. Zeppelin assumed a position *sui generis* in aeronautical history. From their early beginnings to their end, twenty years after his death, virtually all the big rigids ever built were of his design. His name became the name of his invention.

The Count was sixty-two when he completed, in 1900, Luftschiff Zeppelin 1. The idea had been in him for a long time. An entire career as a soldier was behind him. During the American Civil War, as a young German officer in his middle twenties, he had crossed the ocean with letters of introduction to President Lincoln and General Lee. He wished to participate in the war, and he did not care which way. He wanted experience. He called on President Lincoln, offered his assistance, and sought to impress the President by saying that his father before him had been the Graf Zeppelin, and his father before him, and his father before him, and so on. Lincoln told him to try not to let that get in his way. As things turned out, the war failed to interest the Graf. He went West. In Minneapolis, he was given his first ride in a free balloon. He must have liked it. If one balloon could give you a ride like that, he thought, why not a dozen, or fourteen, or sixteen balloons, arranged longitudinally and held within a rigid frame? Placing the lifting gas in multiple cells would help defeat the danger of punctures. In a rigid frame, it might be possible to build an airship four hundred feet long. He did so —thirty-five years later.

Luftschiff Zeppelin 1, 2, 3, and 4 were not uncomplicated successes. They broke, collapsed, and exploded, but they flew. They made unscheduled landings in the Rhine. It was the Graf who flew them, in his white cap, his wing collar, his wicked mustache. People ran through the streets of cities and villages cheering him, above. Church bells followed his flights. He even flew at night. He flew to

Switzerland (the *Schweizerfahrt*), and cruised at forty miles per hour over Lucerne and Zurich. The Kaiser called him "the greatest German of the twentieth century." Zeppelin Platz was dedicated in ceremonies in Berlin. The Graf had apparently solved his early technological problems. In 1909, at the age of seventy-one, he formed his transportation company—with Eckener, whose background was in economics—and began to fly passengers all over Germany. He built rigid airships for use as weapons in the First World War. They terrorized England. Whole populations fled provincial towns. The zeppelins, as they by now were called, sat up in the middle of the English clouds and each one reeled out a small "cloud car" until it hung in clear air, sometimes a thousand feet below. By telephone, the man in the cloud car directed the airship over targets and ordered the release of bombs. A hundred and fifty-seven rigid airships were all that were ever built in the world, and a hundred and thirty-eight of them were built in Germany. The first aircraft to fly across the Atlantic Ocean, an English rigid airship, was a zeppelin copy. The English built sixteen rigids in all. The United States Navy, in the course of time, had four. One of these, the Los Angeles, was a zeppelin built in Friedrichshafen and handed over as war reparations. Another, the Shenandoah, was a zeppelin copy. The Akron and the Macon were of American design. One fell into the Atlantic, the other fell into the Pacific. Meanwhile, under Hugo Eckener, the German Zeppelin Transport Company was continuing the perfect record of its founder, carrying passengers to domestic and foreign cities with no fatalities. In 1928, the company put into service the LZ-127, an extremely slender vessel (its fineness ratio was almost eight to one) that would compile, in the end, by far the most impressive log ever made by an airship, statistical proof for all time of the potentiality of the big rigids. Almost as if they knew how historically important this airship would be, the Germans named it the Graf Zeppelin. It flew one million fifty-three thousand three hundred and ninety-one miles. It once flew around the world, from Friedrichshafen to Tokyo to Lakehurst to Friedrichshafen. It carried, in its nine years, thirteen thousand passengers. It spent seventeen thousand hours in the air. The Graf Zeppelin was only twenty-nine feet shorter than the Hindenburg, and flying together, slowly going in and out of clouds, they were something to see—these phallic, Wagnerian rigids, brushed with mist. Hitler was only mildly inspired by them, but he knew propaganda when it floated by, and he ordered the big airships to present themselves overhead at outdoor rallies and sporting events—parading their immensity and the swastikas on their fins. The airship transportation company was operating at the sufferance of the Air Ministry, so the Hindenburg and the Graf Zeppelin generally appeared as requested, but not always. The company was not affectionate toward the government. Eventually, Eckener defied the Führer once too often, and he was removed from command of the Hindenburg, which flew without him to Lakehurst on May 4, 5, and 6, 1937. The Hindenburg burned while mooring there—a billowing holocaustal white fire. The Hindenburg completely burned up in thirty-four seconds. There was nothing left but fine gray ash and soft aluminum. The huge structure had crumpled instantly into tumbleweed heaps on the ground. The hydrogen burned at about three thousand degrees Fahrenheit. No one watching could imagine the possibility of there being survivors. Luck, though, seemed to run in variegated patterns under the flame. Some passengers and crewmen died at once, but the heat was driving upward, of course, and as the ship crashed the passengers were brought to ground level, where, even within the surrounding fire, there were cooler patches, avenues of escape. Some people came walking eerily out of the flames and fell dead. Others came out with minor burns. A number were completely unhurt. Sixty-one of ninety-seven survived. Thirteen of the dead were passengers—the only passenger fatalities in the history of commercial airships. No satisfactory explanation of the source of the fire ever emerged. It was a stormy day in New Jersey. Static electricity or some form of lightning were the official guesses. A reasonable theory has been advanced that a German saboteur, an anti-Nazi member of the crew, wishing to destroy this symbol of German power, blew it up with a flashbulb. The Graf Zeppelin, on the way home from Rio de Janeiro, was approaching the Canary Islands at the time. The radio operator picked up the news, and the captain decided not to inform the passengers. Their voyage was the last voyage of the rigid airships. The Air Ministry shut down Deutsche Zeppelin-Reederei G.m.b.H. All schedules were cancelled. The ships were, before long, disassembled. At the height of their development, the height of their performance, the rigid airships disappeared.

All that was still ten months away, though, when the Hindenburg flew east with Dollfus, Schmeling, and Professor Labatut. The worst thing that happened to Labatut was that he ran out of film. Looking down at the ocean the second noon, he saw a phenomenon he had never seen (but one that was apparently common in airship travel): two concentric rainbows, complete and perfect circles, were moving with, and framing, the Hindenburg's shadow. The nose of the airship seemed to nudge the inner rainbow. Labatut would have given almost anything for twenty feet of film. Instead, sipping Rhine wine, he rapidly sketched in charcoal and pastels the rainbows, the shadow, a freighter on the sea, and, in the foreground, on the windowsill, the glass that held his wine. (The picture still hangs in his living room in Princeton.) Cattle on the Isle of Man backed up before the big, low-flying airship, and kept walking backward, eyes agog, and Labatut said to Dollfus and Eckener, "Do you know what they are saying? They are saying, 'What a bull!'" A bicyclist on the Isle of Man looked up at the airship as he pedalled along, and kept looking up, and dived into a hedge. R.A.F. planes rose and snapped pictures of the Hindenburg while the Hindenburg made pictures of Manchester and Leeds. Across the Channel, Utrecht was celebrating some sort of centennial, and people ran through the streets waving orange pennons at the Hindenburg. In the slanting afternoon light, the Dutch canals appeared as dark as obsidian. The reflection of the airship in the canals was silver. The Hindenburg was so big that segments of it shone from several canals at once, and its reflection moved from canal to parallel canal like a shuttle through a loom. The Rhine, its improbable turrets reaching up to the improbable Hindenburg, formed the airship's final glide path. Turning left at the confluence of the Main, the Hindenburg dropped mooring ropes over the Rhine-Main World Airport, Frankfurt, at four in the afternoon, June 26, 1936. That voyage, to Labatut, was the sum of the art of flying, expressed in its mild speed, its aerostatic firmness, and its proximity to the earth. The airships, however briefly, had brought into the milieu of commercial transportation what was otherwise possible only in a free balloon. The explosion at Lakehurst ten months later gave Professor Labatut no second thoughts. "I have never regretted for one moment to have had the chance to cross the ocean in the dirigible."

As long as people were alive who had known the rigid airships, the wish would continue that one day they would return. Mercer County Airport may have seemed an unlikely site for the revival, but New Jersey was where the rigids had been; and just as German airship men were living it out in places like Friedrichshafen, Frankfurt, and Zeppelinheim, American airship men were growing old with the wallpapers of, among other places, Trenton. There was in New Jersey a climate of support. Headlong zeal for airships was understood there. One day in the middle nineteen-sixties, William Miller walked into an investment-brokerage firm on Nassau Street, in Princeton. Miller was minding his own securities, which were then voluminous and diverse. As it happened, he was in the midst of a troubled and ambiguous era in his life. He was thirty-nine. His highly developed sense of mission was seeking, and not satisfactorily finding, a sense of direc-

tion. Princeton Theological Seminary had recently con-
ferred on him the degree of Master of Theology, and he
was engaged in various kinds of church work, but he was
not sure he wanted to undertake a formal ministry. Lonely,
self-sufficient less by choice than by force of habit, Presbyterian, he felt compelled to serve God and man but
could not say to himself how best to do it. He watched the
pari-mutuel for a while, the slide show from the New
York Stock Exchange, numbers, letters. In that yeasty
market, his portfolio swelled before his eyes. His interest
in it was less than consuming, though. He literally did not
know what to do with his money or with himself. His
glance fell on a picture that was on a broker's desk—a
romantic sketch, photographically reproduced, of a triple-
hulled dirigible soaring into a mottled, darkling sky. On
the side of the dirigible was the word "AEREON."

"What on earth is that?" Miller asked the broker.

"That? Oh, that's Monroe Drew's airship." Drew, whose
circles of contact were widening, had spoken at a lunch-
eon of the Nassau Club and afterward had distributed the
sketches to rich Princetonians. The broker explained that
it was Drew's apparent intention to take up where the
zeppelins had left off, with the emphasis this time on air-
freight. As soon as it could be arranged, Miller and the
broker went to Mercer County Airport. Miller saw the airship, admired its three cylindroid silver hulls, and met the
Reverend Mr. Drew. God had not insisted that only trained
Presbyterian clergymen or their sons could be the custo-
dians of the Aereons. Coincidence merely made things
appear that way. Miller at once felt the excitement and
saw the potentialities in revival of rigid airships. Before

very long, he would be describing Drew as "a leech, a se-
ducer, elusive and slippery, a dexterous man, who be-
witches people with half a shadow of reality and is able to
weave fact and insinuation in such a way that if things
come out badly it was you who misinterpreted but if they
come out well it was he who guided you." Miller began
their relationship, however, by providing Drew with
money to advance his dream.

Other people were journeying to Mercer County Airport, too, and from places a great deal more distant than
Princeton. A cattle rancher in upland Peru somehow heard
of Aereon and made a trip to Trenton to see if the company could solve his problem. On the way to Lima, his
cattle were dying in the Andes, in a certain impassable
pass, at twelve thousand feet. Could Aereons carry the
cattle? *Sí, hombre. ¿Por qué no?* Production would begin
soon—as soon as possible after the first flight test. Connel-
lan Airways, of Alice Springs, Australia, approached
Aereon. Huge cargo carriers, able to land on patches of
grass, seemed appropriate for the development of the Aus-
tralian bush. Right you are, mate; queue here. These were
light-headed days. "Envision the most luxurious cruise
liner, and we'll match it," Drew liked to tell people. Aereons could go anywhere. They could hover over cities as
flying night clubs with dancing under the stars. They
could lower tourists into the streets of Benares and reel
them in again at dusk for defrosted American dinners in
air-conditioned comfort. "Primarily, though, this can be
the workhorse of the world, literally the workhorse of the
world," Drew would say. "Undeveloped civilizations can
leapfrog from the donkey trail to the Aereon city."

The *Wall Street Journal* sent Robert E. Dallos, staff reporter, to Trenton, and the result was a front-page story on September 20, 1965:

FIRM OF EX-NAVY MEN BUILDS CARGO-HAULERS OF FUTURE: DIRIGIBLES
Ship Will Carry 100-Ton Loads, They Claim; Skeptics Say Plan's Future Could Be Dim

It's a bird. It's a plane. It's a super-dirigible! A group of Navy veterans plans to make huge, three-sectioned dirigibles for use as cargo carriers, bringing what they call "some startling changes" in the air-freight industry. The craft, each of which will be able to carry up to 100 tons and travel 150 miles an hour, will be able to haul containerized freight, huge missile components, prefabricated bridges, communications towers, buildings and oil rigs, the developers claim.

The initial launch of the experimental model is scheduled in six weeks, and, if all goes well, full-scale production will begin shortly thereafter with commercial use beginning in 1967, say the owners of Aereon Corp., the company they formed to build the dirigibles.

In Love with a Lost Cause?

Success, however, is not assured, and skeptics abound. "I admire these men," says an official of the National Aeronautics and Space Administration, at Huntsville, Alabama. But he adds: "Once a man is involved with L.T.A. (lighter-than-air) transportation, he stays with it. But he may be in love with a lost cause." Nevertheless, "if Aereon demonstrates its proven capability NASA would really be interested. It has potential," the official says.

Edward MacCutcheon, head of the office of research and development of the Maritime Administration, says Aereon "is developing some fascinating concepts. This could be a versatile aircraft." He says the "predicted performance capabilities and the economies promised by Aereon's venture are of interest to us as an adjunct to merchant shipping." Aereon sees its dirigibles loading and unloading merchant vessels at sea, saving time and cutting port fees and manpower costs.

Aereon also maintains its dirigibles will be able to haul freight across the country faster than trucks and at competitive rates. And it says its craft will be able to handle heavier loads for less money than existing jet freighters while landing at airports inaccessible to jets.

FEARS FROM THE PAST

If dirigibles are such wonder transports, why hasn't someone thought of using them before? Well, for one thing, the mention of dirigibles still instills fright in many people who remember a series of tragedies in the 1930s, notably the explosion of the airship Hindenburg May 6, 1937. The accident killed 36 persons and all but ended commercial use of . . .

All kinds of people were suddenly interested in Aereon. A Canadian holding corporation wanted a vehicle that could land many tons on the tundra. The Kennecott Copper Corporation had found copper at drill sites beyond the Arctic Circle. Could Aereon bring out the ore? A garbage contractor who regularly dumped bargefuls of garbage in the sea off New Jersey wanted to ease his work with a flying garbage scow. A man travelled to Trenton from western Ireland to ask Aereon to come carry his mutton to market. Drew told the Irishman that his request was somewhat premature. The prototype had not yet flown. The Irishman was broke. Aereon contributed to his return fare, on Aer Lingus.

The company was not only hovering shy of its objective. There had long since developed within the corporate structure schismatic vibrations over what in fact the objective was intended to be. Drew had not named his company Aereon for nothing. He wanted his airships to fly, at least partly, on what he referred to as "G-power"—gravitational power, nature gas, the method of locomotion used by Solomon Andrews for the original Aereons a century earlier. Drew happened to be the editor of the magazine *Military Chaplain,* and, as such, went to Washington twice a month. There he had made his own patent search, and had found what he was looking for: "S. Andrews, Aerostat, Patented July 5, 1864, No. 43,449. To all whom it may concern, be it known that I, Solomon Andrews, of Perth Amboy, in the County of Middlesex, in the State of New Jersey, have invented a mode by which the air may be navigated, and a new and useful machine by which it may be done, which machine I call an 'Aereon'; and I do hereby declare that the following is a full, clear, and exact description of the construction and operation of the same. . . ." Andrews had written a tight and technical document upward of four thousand words long, and with it Drew intended to replace the internal-combustion engine. At the very least, he intended to use gravitational propulsion in supplement to fossil fuels. Drew's enthusiasm was not fully shared by Aereon's first engineer. "The Andrews principle proved to be the curse of the project," John Fitzpatrick said later, speaking ex cathedra from his gas station in Pennsylvania. "Drew had no scientific knowledge. Technology was all a big wonderful world, in his view. He built models out of Prell bottles and talked about 'G-power,' a term he coined. This embarrassed me." Drew had also traced the whereabouts of Solomon Andrews' direct descendants—among them a stockbroker, an ornithologist, an engineer. He put one of them on his board of directors. He told them he thought their ancestor was "one of the worst-cheated inventors in history." Members of the clan invested a hundred thousand dollars in Aereon. It signified much to Drew that Andrews had got his idea while in church. "James Watt credited the steam engine to a devotional inspiration," Drew would point out. "Watt was contemplating the power of God in nature. Solomon Andrews was the son of a Presbyterian minister. He got *his* idea in a devotional situation—in church, during one of his father's sermons. He always considered this a direct inspiration. He had mystical drive. He had twenty-four inventions, but he always considered the Aereons to be the purpose of his life, his real destiny. He had solved the mystery of natural flight. I would like to build an Aereon with no engine at all. It would get into the air on hot air or on helium. It would be flown by controlling the center of gravity. John Fitzpatrick's disinclination to use the Andrews principle is almost laughable, if it were not to me so tragic."

"I regarded Drew as an annoyance," Fitzpatrick would eventually reveal. "Gravitational propulsion was one of fifteen or sixteen things I was testing and experimenting with in trying to achieve complete utilization of the poten-

tials of a combination of lighter-than-air and the aerodynamics of a buoyant wing plus the use of the helicopter
rotor not for lift but for propulsion and control. Solomon
Andrews had no particular influence on what I was doing.
He had an imperfect understanding of fluid flow and aerodynamics in general. I thought he should get proper credit
for what he had done, that's all. I wanted to build an aerobody, but the curse of Solomon Andrews was on the corporation. Drew was pitching his money-raising efforts on
the mystical qualities of the first Aereon. We built the
wrong vehicle, the three-hulled vehicle. We had been
trapped by the Solomon Andrews thing."

"The ship was designed to demonstrate the Andrews
effect," Drew would say. "But because of a slight error on
John's part it could not do so. It weighed three hundred
pounds too much."

While all these pustules were festering, Drew, of course,
had his perennial problem of finding money to keep the
corporation alive. The search eventually led him to the
door of William Sword, an investment banker who also
happened to be a ruling elder in the First Presbyterian
Church of Princeton. Sword, potentially, was the answer
to Drew's prayers—a figure of loft in the laity of a sister
church and, simultaneously, a Wall Street financier, a
partner in Morgan Stanley & Co., whereunder the green
waters ran extremely deep. Sword greeted Drew cordially
and led him into his living room. Sword was an affable
man of pebbly chatter. He was scarcely forty, younger
than Drew had expected he would be. He and Drew had
much in common in addition to their close association with
the Presbyterian Church. Sword's specialty, within his

world, was sales. So was Drew's. Each had put in a long apprenticeship knocking on igloos, and each had become a super-salesman, a megaflack. The two men circled each other verbally—a long theological sniff with commercial undertones. Drew outlined the opportunity: the salvation of the world through big rigid Aereons, formed into a Christian Faith Fleet. Sword excused himself, saying that he was technically unequipped to evaluate this—did Drew mind if Sword called a friend who was an aeronautical engineer? Certainly not. Sword went into his study and called his friend, who said, among other things, "That company is a big joke." When Sword returned to the living room, his eyes were as friendly as two nickels. Packaging the conversation, he showed Drew to the door. The contact might have ended there—just another rebuff in Drew's tireless rounds—but Sword took it upon himself to protect the innocents of Princeton from further approaches by the reverend minister of the Fourth Presbyterian Church of Trenton. According to Sword, one reason he felt this responsibility was that Presbyterians kept coming up to him—widows, deacons, threadbare academics—with the light of hidden provender in their eyes and saying, "I understand *you* know Monroe Drew." "Slightly," Sword would say. "And I frankly don't think he's got anything more than a kite. He is using the church to get his company financed. He is selling stock to pay off loans. The whole thing disgusts me." If Drew kept hounding people in Princeton, Sword said, he was heading for trouble with the securities people. At a coffee hour following a Sunday-morning service, Sword talked to Miller and warned him to look out for his investment. All this filtered back to

Drew, who eventually decided that Sword was a threat to Aereon and had to be silenced. There was no need to go to secular law. Within the codes of the Presbyterian Church itself there were provisions for handling people like this. Drew called the minister of the First Presbyterian Church of Princeton and told him that if Sword did not at once cease to belittle the Aereon Corporation, Drew was going to arraign Sword before the session of the Princeton church, there to answer charges levelled by Drew himself. There were laws of centuries' standing within the Church of Scotland applicable here, Drew said, establishing the form of such an ecclesiastical trial. If Sword was found guilty, he would be condemned to the cutty stool, the low stool of penance, on which he would sit facing the congregation on as many successive Sunday mornings as the plenary session might deem appropriate and commensurate to his offense. Sword turned into a plowshare. According to Drew, not one negative syllable was ever again uttered by Sword about Aereon.

Belial did not always back away so quickly—as when trouble came to Aereon in the form of the Securities and Exchange Commission. There are three ways to float stock: by private offering; publicly, through full registration with the S.E.C.; and, also publicly, under an S.E.C. exemption from full registration, known as Regulation A, which allows companies to issue a maximum of three hundred thousand dollars' worth of stock without incurring the avalanchine legal fees that go with full registration. Early in 1965, Aereon had applied for a Regulation A exemption. It was granted in the spring of 1966, almost on the exact day that Monroe Drew's airship, the triple-hulled Aereon, was blown over by the wind and bulldozed back into the hangar, never to emerge again. That summer, a computer created the deltoid Aereons, and the computer was paid with money from sale of the new stock. Just after the turn of the year, without warning of any kind, the S.E.C. telephoned and told the officers of Aereon to appear in Washington the following afternoon with the company's books. Sale of stock was halted while the S.E.C. gathered facts. The S.E.C. took depositions in Trenton. What or who had triggered this investigation has never been revealed. The strain of it cracked the company to pieces. Board meetings were dipped in savagery. John Fitzpatrick, who, in addition to his engineering functions, had for the past three years been playing president to Drew's founder, quit. "I have long felt that ministers get between man and God," he explained. "Drew has a monstrous ego. He has demonstrated consistently over a long period of time that he is incapable of living with reality. In selling the stock, he talked the moon. Everyone in the organization at one time or another has tried to put a wet blanket over him. In his approach to investors, he has never failed to be Reverend Drew. If you really want to know what I think of Monroe Drew, get a copy of *Moby Dick* and read Father Mapple's sermon. Jonah was a man who tried to flee from God. 'Woe to him whom this world charms from Gospel duty. Woe to him who while preaching to others is himself a castaway.' I wanted to build an airship that could maneuver when landing or taking off, that wouldn't blow away in the first big wind. Meanwhile, Drew was talking about 'G-power,' saying it was 'something mystical.' He was going to revolutionize aviation transportation. He overstated the credentials and qualifications of everything. He embarrassed me until I wanted to slip through a crack in the floor." Drew, for his part, said he was sincerely sorry, and surprised, that Fitzpatrick felt this way. "When we got into trouble with the S.E.C., John turned against me."

The S.E.C. compiled a list of particulars, and the general drift of it was that Aereon had misled its investors in various ways—for example, by not properly informing them that the entire triple-hull concept was being scrapped in favor of the deltoid configuration, by failing to disclose that no engineering data existed to support certain engineering theories, and by using false and misleading sales literature and making false and misleading oral statements, such as the following claims: "the proposed aircraft can deliver four to six times the average truckload, at ranges up to four thousand miles, at a cost of less than one and a half cents a ton-mile; the issuer is proceeding as rapidly as possible toward manufacture of a commercial aircraft; and rapid progress is being made toward a commercial prototype aircraft." All the company actually had at the time was a computer's description of a deltoid rigid airship. John Kukon's flight of a twenty-inch Aereon was still many months away. The company had no president. Its treasurer had also quit, and its other engineer, Jürgen Bock, would soon go home to Germany. About a hundred and fifty thousand dollars had been raised from sale of the Regulation A stock. Most of it had been spent. The S.E.C. permanently suspended the Regulation A exemption, and Aereon could raise no further money by public sale of stock. Lazarus at his worst had had a stronger pulse.

I N A E R O N A U T I C A L C I R C L E S around Princeton, Aereon was viewed with, among other things, cruel amusement.

"I don't think it is generally understood that technological advances are not made by back-yard inventors anymore."

"There could be something about it that no one at present knows. It *could* be beyond the man of ordinary skill in the art. I doubt it."

"The man of ordinary skill in the art used to be Thomas Jefferson inventing a better plow, but now it's a man with a master's degree and twenty years' experience in the business."

"A bright-eyed minister of theology and a gung-ho airship man do not add up to technical competence."

"Oh, you're just one of those theoreticians who can prove a bumblebee can't fly."

"I get a twinge of sadness whenever I think about Aereon."

"Drew is wild. He is a fighter-pilot type. You need them, sometimes. If I had a product to make, I'd want Drew to raise the money. I think he can get blood out of a turnip."

"He should have been a pitchman in a carnival selling patent medicines."

"Fitzpatrick was an energetic, shirt-sleeve, self-trained engineer who got over his head in a technical field in which he was not really qualified."

"No serious student of aerodynamics would believe that that triple-hulled airship was an efficient configuration design. The big question now, with the delta configuration, is: Is this an efficient design for the purposes for which it is intended?"

"You can make anything fly. You can make the George Washington Bridge fly if you put enough engines on it."

"What I was taught is that if you want to pick up a big load off a field in a short distance you build a long, straight, thin wing. A high aspect ratio (span over chord) on a conventional airplane is the way to do this. Aereon may have a fantastic invention here that nobody understands, but I think that is unlikely."

"There is real merit in putting bags of helium into the structure. You can calculate this. There is no mystery about it. For engineering purposes, Newton, on buoyancy, was right. You can calculate whether to extend wings or add helium. Aircraft configuration is not an exact science, but it's somewhere near it. Thus, if this were such a God-damned good idea, it would have emerged long ago."

"It's just part of the American dream. You know—'Your

boy, in a bicycle shop, can invent something that will change the world.' The finest scientific minds in the country told the Wright brothers they couldn't do it."

"Two wrongs make a right. Two Wrights make an airplane."

WILLIAM SWORD, by the oddest of ironies, may have resurrected the Aereon Corporation, if only indirectly, catalytically. For when Sword told William Miller to beware, and to watch his investment, Miller had begun to attend Aereon board meetings, and had changed from a passive investor to an active participant. Slow to commit himself to anything, Miller had a capacity to be a passionate believer, and once he believed in something, he would characteristically devote himself to it with a loyalty and an expenditure of energy that could obscure almost everything else in his life. Miller believed in Aereon. He believed in the word of the computer when it said that the form in which to make a big rigid airship that could carry huge loads and land like an airplane and fly under aerodynamic control was to hunt for the optimum compromise between an airfoil and a sphere. A sphere has the greatest possible volume relative to surface area, but a sphere has too much drag. A sphere is a bluff body as opposed to a fair body. So you elongate it, make an ellipsoid. A conventional airship is an aborted ellipsoid. Keep on flattening it, toward the airfoil. Press down on it. Shape it. Widen it aft. Make it as broad in the stern as it is long. A triangle with a deep belly and a vaulting back should do. No reason the thing should not fly on its own. Helium, however, would add free lift. A deltoid Aereon filled with helium and trailer trucks could fly on sixpence and a song. So the computer said.

Miller believed. Aereon galvanized Miller, and salved his need for a sense of direction. Among all officers and stockholders, he was about the only one left with much of anything that could be called a sense of direction, as far as Aereon was concerned. He was also roughly the only person on earth who at that time had money he was willing to put into Aereon. Surveying this otherwise unpopulated wasteland, Drew and the board asked Miller to become president. He accepted. Some stockholders put up collateral so Aereon could obtain bank loans, but after a time no one was willing to do even that. For more than a year, Miller paid all operating costs himself. His relationship with Drew, warm at first, gradually formed crystals and then froze solid. Miller initially had felt compassionate toward his fellow-theologian, sympathetic as Drew absorbed the causticities of Fitzpatrick and the affronts of the S.E.C., but ultimately Miller grew wary of the founder's overleaping style. It would not do—not in the new and cautious Aereon that Miller wanted to build. The founder sat, as ever, at the boardroom table, but Miller gradually encased him in clear plastic. A showdown at length came, over some minor point of company policy, and in a vote by the shareholders Miller was endorsed and Drew defeated. Drew had become founder non grata, and he was not invited when the tests of the 7 and the 26 were conducted by Miller's consultant engineers.

"Miller would like to wish out of existence the whole previous existence of Aereon," Drew said. "He thinks of himself as the father of the company. He hates to have to refer to me as the founder."

"Drew has his own particular filter of fact and fantasy," said Miller. "Understanding him is like trying to adjust the color on a television set."

"I can still bring into this thing money that doesn't like Miller. Trucking money, as a matter of fact. To get Miller out of the way. For a time I tried to block everything he tried to do. I was a real s.o.b. Then I was quiet for a while. Now I am ready again to attack him. He just will *not* include me in inner management. He won't give me a single picture of the aerobody. He is a great enigma to me. He has a rich boy's education. He has lived as a rich man all his life. He is an unusual brand of churchman—very narrow-gauge."

Aereon was costing as much as fifteen thousand dollars a month. To raise money—his personal fortune having long since run out—Miller had to go into a slow and meticulous waltz with the overreaching shadow of the S.E.C., abiding, and then some, by the rules of private offerings. He could not talk to anyone except "sophisticated and knowledgeable investors." He could not call on a large number of people. He could not call on people over a long period of time, or it might be found that he had made a "continuous offering," which he was not permitted to do. He behaved as if the Securities and Exchange Commission were a guillotine and his neck were permanently on the block. He talked only to sophisticated and knowledgeable millionaires, several of whom came through with money to keep him going. In his approach, he never mentioned Faith Fleets or Christian Freight Lines. He did not suggest that the basic purpose of the development of the Aereons was to advance dramatically the image of God.

I once found myself wondering, though, how Miller might articulate his deeper purposes, if he would agree to do so at all. Personal profit could hardly be the radical motivation of a man who had poured away three hundred thousand dollars without apparent qualm. Monklike, frugal, denying himself every sweet in the box, he had spent (by then) several years singlemindedly struggling to preserve this checkered company. I asked him, around 4 A.M. one night on the way to a flight test at NAFEC, what he saw as his ultimate goal. He sighed unhappily and said he had scruples about keeping his private aims separate from his corporate objectives. He said, "However, if people probe me as to why I'm doing this, I'll tell them, and it always reaches to ultimate values—but I don't ordinarily go into it. The Faith Fleet was not a theological matter but a straightforward sales device—a gimmick. Should the Body of Christ run schools, highway departments, police forces, transportation companies? I don't think so. Christians are supposed to be yeast and salt, the church in the world, people among people. The computer in Valley Forge was a tool created by man, not by God, and the computer created the aerobody, which is also a thing—a neutral object. It could be used for anything. The aerobody itself cannot be good or bad. It's the use of it that can be bad. That is what sin is all about. Like any other gift, an aircraft can benefit man. I see the theological dimension of the aerobody as bringing a means of transportation of general benefit to man. The aerobody would benefit more people than high-speed aircraft do. An executive jet isn't something a small town in Ghana could afford. But unit cost and utilization possibilities would make it possible for a small town in Ghana to afford an Aereon. Could Nigeria afford an SST? No. But Nigeria could use a flexible means of transport carrying many people and a great deal of produce cheaply. We can give developing countries a chance to become more effective in trade. We don't want people to be dependent on us. This is a product they can buy and use in their own commerce. They need no fixed capital investment in a right-of-way. There are broad social concerns, you see, in the national interest—good for the free world."

Miller was so absorbed in what he was saying that he almost negotiated a traffic circle on the diameter rather than the circumference. Fighting the wheel, plunging on through the darkness, he continued, "I want to be a good steward of what has been entrusted to me in Aereon. I give it my life, my time, my money (money is life translated into exchangeable value), my abilities. I'm not the possessor but the trustee of these things. I see purposefulness in investing time, money, and ability as God has given them to me. I don't see meaninglessness. I assume there is a point to life: discovering God's purposes in one's life. The ultimate value spot is God himself. If you have an ultimate value short of that, you have an idol. Paul said, 'I know how to be abased and how to abound.' In Aereon, we sure have been abased. I wonder how we would do if we were to abound. We would have a new means of transportation, and we would have seen the project through a period of great technical skepticism. The basic scheme is to produce a three-hundred-and-forty-foot aircraft capable of enormous lift. Then, later, at the thousand-foot length, we could replace ocean freighters. If the project succeeds financially, I will feel grateful because the Aereons will produce money that I, in turn, will be called upon to administer. Stewardship, faithfulness, quality, integrity, hope, responsibility, trust—these are key concepts with me. I would like to give the profits of Aereon to Christian agencies that serve the Body of Christ—for example, to the Scripture Union, an agency that promotes the strengthening of the infrastructure of the church through daily input in people's lives of meditation on the Scriptures. I would also like to create scholarships for theological students in other countries. I would also like to buy books for theological libraries abroad. And I would like to give as much as I can to the American Bible Society. If I ever get married and have children, I would plan to pass on this stewardship to them. If the company outdistances my talents, I'll try to be a good steward, and not be a possessor. No one is infallible. Meanwhile, to break new ground—to break new air, let us say—is possible because of hope. The reason for my hope is the Resurrection. If the Resurrection is true, and if God does create life out of a dead situation, the Resurrection is God's ultimate solution to man's problem of sin. With God, all things are possible. It is possible that, despite your own frailties and weaknesses, anything can be achieved—if this is God's purpose. God has created us as creators. God has created us as responsible to our brothers. I have been called to this now. If this project fails, I will existentially be questioning my stewardship. If the project fails, though, I know that I will have something for which to give thanks—for character development, if nothing else. I couldn't enter into a vocation unless I felt it was something that God wanted me to do. In this sense, I think of it as a calling, a form of obedience to God. If the air were full of gigantic machines, it might not be good, but, on the other hand, it would be a relief from congestion on the ground. As a Christian, I feel a sense of mission about this. Not everyone in the company shares my trust in Christ."

This was true enough. Some of the others actually worried about Miller's religion. One of them had once said to me, "No one who folds his hands and says prayers at a business-lunch table is ever going to raise enough money to keep Aereon going."

I asked Miller why he did not have a church, since he was qualified to be an ordained minister, and why, instead, he had chosen such a novel form of mission. "I went to divinity school to learn how to help people spiritually," he said. "And not necessarily in a church. All my life, people had said, 'Are you going to be a missionary like your daddy?' My dilemma was: If I did what I wanted, he wouldn't be happy, and if I did what he wanted, I wouldn't be happy."

When Miller was ten years old, in 1936, and his family was in the United States on furlough from Iran, he wrote to T.W.A. and suggested that they build a chain of floating

airports, like lily pads, across the Atlantic—a concept that
Rutgers University would advance, with concomitant pub-
licity, in 1966. In military school, in Chattanooga, in 1941,
Miller frequently daydreamed about an all-freight airline
with him as founder and president. At Choate School, in
1943, he got interested in rockets and helicopters. He
thought out a way to overcome torque in helicopter rotors:
hot exhaust channelled through the blades and out the
tips as jets. Sikorsky Aircraft finally did just that in 1965.
At Choate, he also did an engineering drawing of a projectile that could be fired from a gun and then turn into a
rocket. The projectile was wrapped in plastic containers
holding liquid fuels. The spin of the projectile would force
liquid oxygen or liquid hydrogen out of plastic sacs and
through tubes in the axis to a chamber at the rear. The
liquid fuel would combust in the chamber, blow out a
plug, and turn the projectile into a rocket. He actually
explored getting a patent. In 1944, aged eighteen, he went
into naval aviation—Patuxent, Olathe, Ottumwa, Pensacola.
He shot colored bullets through banners. He paused before
recruiting posters that showed Navy pilots over the words
"Rough," "Tough," and "Smart." He found this "ego-
building." He became an award-winning dive-bomber.
"Pointed at the ground, you know, you pick up tremendous
speed. We dived from twenty thousand feet at an eighty-
degree angle. It felt perpendicular. Winds would vary at
different altitudes. You released your bomb at twenty-five
hundred feet, then pulled out at a thousand feet. I lost one
of my good friends that way—really a terrific fellow. After-
wards, I circled around and saw nicks in the ground
that had been made by his prop—he had been that close

to pulling out of the dive. That was a shock." At the end of his Navy experience, Miller was flying Phantom jets in the Reserve. The first place he ever landed one was at NAFEC. He was in college by then, an overage undergraduate in the Class of 1953 at Princeton. If anyone had ever wanted to be an aeronautical engineer, it was Miller, but he felt inexorably drawn to a higher calling. "I thought I was a Christian, but I didn't have the Peace of God. I prayed, 'God, show me more of your truth.'" God answered this prayer at the Campus in the Woods, an enclave of the Inter-Varsity Christian Fellowship, on Fairview Island, in the Lake of Bays, Ontario. Having felt that he was only a "nominal Christian," who had "no real fellowship with Christ," Miller went to Fairview Island to get it. The Campus in the Woods was a nest of fundamentalism. Miller at first looked upon the others there as "people who couldn't make it socially and had collected for solace." They sang hymns and talked about Christ all the time and, in his own word, disgusted him. They had an air of smug certainty about their relationship with God and what He wanted them to do. Miller found this presumptuous. In Iran, Miller's father was praying for a positive outcome of his son's experience in Canada. Lecturers at the Campus in the Woods explained that Christ is a bridge on which man can cross the chasm between himself and God. "God forgives," they said. "And Christ died not only for the sins of the whole world but for *your* sins." Miller found himself becoming interested. "It all built up in three days. Someone said to me, 'By the way, Bill, would you be willing to do God's will in your life?' I said, 'Yes.' I realized I'd said something pretty momentous. The hymns began to

mean something. It all got to me. I no longer felt a di-
chotomy between my will and God's will. I felt, 'He will put
into me a desire to do what I must do—whatever it is.' In
the third chapter of John, Nicodemus comes to Jesus at
night. Christ says to him, 'Unless a man is born again, he
cannot see the Kingdom of God.' That was my autobiog-
raphy, written two thousand years ago. Now my life became Technicolor. Before, it had all been shades of gray.
A letter came from my father saying that he had been
praying that this would happen. Wow! Fantastic! A clear
case of how God answers prayer! By Grace you are saved,
through Faith." Back at Princeton, he spent a high per-
centage of his time trying to encourage his classmates to
follow Christ. He organized eleven Bible-study groups—
antagonizing the Presbyterian chaplain, because the groups
had not been organized under the chaplain's aegis. He
considered flunking out so he could proselytize full time,
but decided at the zero hour that "flunking out would be
bad witness to the faculty," so he took and passed his
examinations. He also worked as campus recruiter for
naval aviation. He combined aviation and religion in his
persuasive approach. When he found a prospective jet
fighter pilot, he would tell him, "What Christ means to me
is something far more wonderful even than flying." For
the History Department, he wrote his senior thesis on
"The Origins, Manifestations, and Products of Presbyterian
Missionary Efforts in the United States, 1789–1837." In
1956, Miller was the president of the student body at the
Biblical Seminary, on East Forty-ninth Street, in New
York, becoming a Bachelor of Sacred Theology. For the
rest of that year, he studied the stock market. Then, for

four years, he worked with foreign students at Columbia University. He was not part of the university staff but operated out of a small office on Morningside Heights paid for by the Inter-Varsity Christian Fellowship. His aim was to enable "friendships in depth" to arise between foreign students and commuters. Miller defined commuters as "Christians who live in suburbia." He went out into "the hinterlands" and scouted, and found a hundred and fifty families ready for friendship. The Protestant chaplain of Columbia cried foul, because Miller was not operating under his aegis. Miller characterized the chaplain as "a Protestant oaf." The chaplain said Miller was insufficiently ecumenical, because he did not seek out Hindu or Muslim families. "Gosh," Miller said. "I had Baptists, Lutherans, Plymouth Brethren—but quality control was everything. I couldn't control the quality of a Hindu or Muslim family, because their outlook was so very different." Miller began to see himself as a revolutionary forever fighting the establishment in the form of academic chaplains. He would feel that way about his struggle for the survival and growth of Aereon. He felt that way about his adviser while he was working for his master's degree at Princeton Theological Seminary. He described his adviser as a Marxist theologian who believed that structures must be taken down, that the status quo is evil, and that sociopolitical action was where the Kingdom of God was at. The professor looked upon Miller as a nineteenth-century missionary. He denied him admission to the Th.D. program and thrust upon him a terminal master's. Not long thereafter, Miller saw the sketch of the three-hulled Aereon and met the Reverend Mr. Monroe Drew.

G OLDSTEIN, in the nineteen-twenties, solved the differential equation for the flow field in the wake of a propeller, and Goldstein was ignored for twenty-five years. For water, as for air, propeller design before Goldstein had been entirely empirical, and it remained almost wholly so until the advent of the big computers, when people were finally able to figure out what Goldstein was trying to tell them. Theodorsen embraced Goldstein's theory, and expanded it. Rose, in his turn, added touches of his own, and assumed preëminence among propeller designers in the United States.

In the most efficient of propellers, the vortex sheets move back as if they were solid. Behind Aereon 26's lemonwood prop, the vortex sheets were apparently moving back in shreds. Hence, the 26 could get off the ground but not much farther into the air. Perhaps Henry Rose could solve the problem. Rose could offer no guarantees, but he could at least give Aereon a propeller that was exactly designed. It would cost five hundred dollars—payable to Rose's employers, the Sensenich Corporation, of Lititz, Pennsylvania. A compound of low buildings beside Lancaster Airport, Sensenich looked like an old lumberyard that had seen its boom. A man from Sensenich was on the road all the time, in northern New York and in Canada, hunting birch and maple. Wood that grew in cold air was particularly hard. The wood was dried in Pennsylvania until ninety-three per cent of its moisture was gone. Then it was cut, planed, and studied—for defects, for grain angle, for grain straightness. Half was rejected.

It was Henry Rose's decision that Aereon's propeller should be made of Canadian yellow birch, and that it should be four-bladed. Rose was limited to a forty-eight-inch diameter, because the engine was so light and small. Most light-airplane propellers were half again as large. This one would be a miniature, more or less, of the big four-bladed props that pushed the Hindenburg. There was nothing conscious in the choice. Rose had no idea what the Aereon Corporation was about. He was curious, but had not been told. He had been given only the essential facts—of power, of weight, of engine rotational speed, of the vehicle's design speed. Air flows over a propeller in much the way that it flows over a wing, with the difference that the propeller moves variably faster from hub to tip. Rose, out of Goldstein by Theodorsen, had to find the optimum blade-load distribution.

The propeller blank was a wooden cross made of birch laminations held together by resorcinol glue, the only glue officially rated waterproof. Specimen propellers had been left on the Sensenich roof for years and years, rotting away, until all that was left was the resorcinol glue. Marks made at intervals on the wooden blank were known as Stations Along the Blade. For each station, Rose had designed a brass template. Carving had to conform precisely to the templates. There were hundreds of carvers at Sensenich during the Second World War. There were two now—tall middle-aged men with glasses, Mel Eichelberger and Edwin Miller. Metal had long since become the material of most propellers. But wood vibrated less and never had fatigue. Metal propellers were more efficient, but for some things metal would not do. Metal required forging and dies. For experimental purposes, no one could afford such a die. Eichelberger and Miller looked like cobblers. They carved by hand. They used drawing knives, with two handles, and powered abrasive drums. They carved by feel. They felt the fairing of the blade. And they were watched by Henry Rose, a kindly, gentle, and scholarly man, tall, with dark-rimmed glasses, and an insignia of rouge in his cheeks.

Rose had grown up in this milieu, and had first met the Sensenich brothers forty years before. The brothers, as boys, had carved an empirical propeller from a walnut bedstead, attached it to a motorcycle engine, attached that to a sled, and skimmed across the snow down the mile-long lane from their family's farmhouse to their mailbox at the county road, returning with the *Saturday Evening Post*. They built their snowmobile because they hated trudging for the mail. They whirred along on the ice of the Susquehanna River. They carved new props for Jennies from the First World War, because there were no surplus props. The Sensenichs were Mennonites, out of step with custom. Their people went along the roads of rural Pennsylvania in horse-drawn black carriages, ignoring automobiles, rejecting such aspects of the century, while the brothers Harry and Martin Sensenich made fortunes in aeronautics. God's ways were hard on their family. Their father was killed by a train. Their mother died in a fire in her kitchen. Their brother was kicked to death by mules. In 1944, the Sensenichs carved forty-three thousand propellers.

The tapering blades of Aereon's propeller, as they took shape, spiralled with the grace of lanceolate leaves. The templates fit perfectly the Stations Along the Blade. The propeller was dipped in sealer and sprayed with varnish. The leading edges were trimmed with brass, which was riveted into the wood with copper. Two coats of polyurethane covered all. The grain shone through. The propeller was ready. A work of mathematics, not of art, it seemed almost too beautiful to hang on a wall. Rose sent it off to New Jersey, wondering (a single, pusher propeller? so small?) what on earth it was for.

THE PROPELLER was ready but Aereon was not, since Miller had scarcely enough money to pay Sensenich and none at all for renewed tests. His fund-raising efforts failed totally. The market environment was bad, he explained. The 1970 year-end sell-off was begin-ning. Christmas was a head wind in itself, against which he could not make progress. His consultants were drifting into other work, but Miller refused to go on with the tests until he had what he called "funds identified in hand." With the 26 hidden away again in its Sheetrock box at Red Lion Airport, forty miles from NAFEC, Aereon went into hibernation. John Olcott, the test pilot, speaking to Miller on the telephone, said to him, "I just sometimes wonder if you really want to see the final results of these tests."

"How can you say that when you know I've put all these

years and all my own money into this project?" Miller said.

Olcott said, "I don't know. I'm just wondering, that's all."

Miller went off to Urbana, Illinois, to represent the Scripture Union at a triennial convention of the Inter-Varsity Christian Fellowship, the purpose being to persuade twelve thousand assembled students to be foreign missionaries. Miller had not missed one of these convocations since he was in college. Back in New Jersey, he continued the hunt for money. Budgeting pessimistically in terms of weather and time, he figured he needed fifty thousand dollars for just enough additional flight tests to show, one way or the other, the pertinent validity of the Aereon concept. Many weeks went by. Three existing shareholders finally gave him forty-six thousand dollars. It would do.

February 24, 1971, about half an hour before dawn, Buddy Allen, of Tabernacle Township, who was getting a hundred dollars for the job, started to move his flatbed trailer out of Red Lion. Aereon 26 was chained down to the bed and camouflaged as before, with newspapers taped over all its markings, more newspapers over the canopy, and a tarpaulin shrouding the engine and the prop. The light in the east was medium gray. An acre of now frozen mud around the apron of the little hangar had been striated with tire ruts, and the ruts were white with panes of thin ice. The big truck cracked the ice as if it were driving slowly over bottles. It circled a biplane and lurched up onto a blacktop road.

A state-police car led the way, followed by Everett

Linkenhoker's station wagon with a big sign on the roof—
"WIDE LOAD." Then came the 26, its wide rear reaching
out on either side beyond the macadam; then Miller, in
his Mercedes (I rode with him); and, finally, Paul Shein
and John Weber, with another roof sign—"OVERSIZE LOAD."
Miller remarked that the procession reminded him of a
state funeral. A fin hit a juniper bush. The corresponding
fin on the other side hit a branch. The first fin just missed
a road sign, skinned a mailbox, and barely missed poles
and trees, more poles, more palisade trees. The amperes
were rising in Miller's nerves. The Mercedes' radio was on,
and a concerto was coming out of it. Miller snapped it off.
He felt guilty. "I don't like to shut off classical music," he
said. "It somehow seems impolite."

The road widened out into Red Lion Circle, and the
procession began to move south on Route 206, a wider
highway, but still two-lane, with the 26 reaching far
out over the center line and far out over the right shoulder.
Dawn came. The state trooper waved farewell. Another
trooper would meet the 26 near Hammonton.

"This is great," Miller said. "Wintertime. No one is out
wandering around in their yards in the early morning."
A pickup truck passed the 26, going down the far shoulder.
A blue Ford and a small red van followed the pickup,
then various automobiles. Miller watched them all suspi-
ciously. Anything with a long antenna made him nervous.
He could see the enemy and the curious coming in in
pincers: Boeing, Douglas, Lockheed, *The New York
Times*, the Cleveland *Plain Dealer*, the New Orleans
Times-Picayune. The procession was moving at fifteen
miles an hour. Two big tractor-trailers and several other

vehicles were bunched up behind. Huge trucks were now coming the other way, raging north, pressing up the long straightaways at seventy miles an hour, dropping six wheels onto the wet shoulder and spraying sheets of water as they rocked the 26 with blasts of wind. Some of the big diesels missed the aerobody's anhedral by inches. One slight contact and the 26 could be ripped in half. Linkenhoker, in front, moved his station wagon over a little more into the path of the trucks. He all but stuck his left front fender into them. The drivers leaned out and shook their fists. They were on their way to Whippany, or wherever, and they did not like the wide load coming into them. They could hardly have known or imagined that the ambition of the 26 was to grow large enough to swallow them and their trucks and carry them through the sky.

After a time, the big diesels that were following the 26 made their move and passed, going off the road on the far side. Their sudden absence in Miller's rearview mirror revealed a brown Plymouth sedan. Miller kept an eye on it for a while, then became alarmed. The Plymouth was making no attempt to pass, even though the road ahead was straight and free. "We've got a follower," Miller said. "He's lagging way behind, so other people can get by him. He definitely does not want to pass us. What is his plate number?"

"KMN 260."

"Perhaps we should write that down."

KMN 260 followed the 26 for several miles and then passed, and then stayed in front for a while, pacing the procession, before moving on out of sight. "I imagine he took pictures," Miller said, with a resigned sigh. Miller seemed to accept this calmly, with experienced stoicism. The cortege moved on, now doing twenty miles an hour. Miller continued to watch closely the road, the traffic, the landscape. Blueberry fields came up on the left, and a peach orchard on the right. His glance searched through the orchard as if there were transmitters hanging from the branches of the trees. The 26 approached Hammonton Circle. A state trooper was waiting there, on foot, in a forest of warning signs. He had cleared out all traffic, and he directed the 26 to go through the circle clockwise, a more direct route than the orthodox one, and less endangered by the razor-edged signs. Passing him, Miller gave the trooper a military salute. The trooper returned it. The procession headed east toward NAFEC, on the White Horse Pike. The road was now larger, and more popular—four lanes, undivided. The 26 filled its side, but did not cross the center line. Traffic amassed behind—ten, eleven, fifteen cars. Miller shuddered. The third car back was a brown Plymouth sedan—KMN 260. Perhaps the driver had gone off to mail pictures to Seattle and had now come back for more. So intent was Miller on the Plymouth in the mirror that he nearly plowed into the 26. The procession had now grown to some twenty-five vehicles, and there were signs of impatience, scattered hornplay. Suddenly, Pennsylvania 37C 109 had had all of this that Pennsylvania 37C 109 was going to take, and Pennsylvania 37C 109 leaped out of the pack, crossed the double yellow line, vroomed past the 26, and shot away. KMN 260 also jumped out of the pack and went around the 26 like a flicker of light, chasing Pennsylvania 37C 109 in a race that lasted fifteen seconds and ended with the

Pennsylvania car stuffed into the roadside, its driver under arrest. KMN 260 was an unmarked patrol car of the New Jersey State Police. Its occupant, who was having a morning and a half, was the same efficient fellow whom Miller had saluted only minutes before. At the first sizable turnout, an abandoned drive-in, Buddy Allen removed the Aereon from the highway and loosened up the traffic. Moments later, the trooper pulled up beside the 26. He got out of his car, snapped shut his summons case, and, with a booming laugh, inspected the aircraft.

"That thing ain't going to fly," he said. "That thing ain't go-ing to fly. What is it, anyway?"

"It's an aerobody," Miller told him.

"It looks like an orange," said the trooper, and, with another tympanic laugh, he drove away.

As the motorcade moved on, I watched the faces of the people in oncoming automobiles. Almost no one noticed the 26. It could have been a dome mobile home or an eleven-hundred-pound pizza. It could have been a flying saucer or a lithic slice of the moon. It did not seem to matter. I did see two jaws drop. While Miller was absorbed with a suspicious green Pontiac, a small white car passed the 26 on the right and nearly clipped off a fin. The young woman in the car raised a finger. Miller said, "She probably thinks we're part of the military-industrial complex." A NAFEC police car was waiting. Its siren began to wail. It led the 26 through the NAFEC gates and across the two final miles to the big hangar. "They'll only give us three weeks maximum here," Miller said as he parked. "NAFEC has become difficult, skeptical, and impatient."

The NAFEC policeman got out of his car, sighed, chuckled, and said, "You'd never get me in that son of a bitch. You'd have to tie me."

At least thirty people had surrounded the 26 in wonder on its first arrival at NAFEC. Now one heavy-lidded man in white coveralls shuffled over and stood by the NAFEC policeman. "They never got that thing more than fifty feet off the deck," he said. "It never turned to the left. It never turned to the right. The pilot just flew straight and low, over Runway 13."

"I don't blame the son of a bitch," said the policeman.

Sheets of the Trenton *Times* were peeled away from the fuselage: "OUTLOOK BRIGHT FOR YULE BONUS." "SECOND CHALICE STOLEN, RETURNED TO CHURCH." As the new propeller was uncovered, the man in white coveralls whistled. Something interesting had happened after all. "Look at that," he said. "Look at that propeller. That thing must have cost six thousand dollars if it cost a cent."

O N   T H E   F I R S T   D A Y of March, the tests re-
sumed. It was a mild, calm morning. At dawn, the
26 was already addressing itself to the big runway, and
had been cleared by the NAFEC tower. This ultimate
phase of the test program had been coded the Quick
Look. The test outline included seven "tasks," designed to
yield the "hard facts." Miller and the Aereon board of
directors could then decide, and would have to decide,
whether to liquidate their company, after twelve fruit-
less years, or to move on toward the building of the big
rigid Aereons. On the day's first run, Olcott never left the
ground. This was intentional. Task 1 was "high-speed taxi
without rotation." Its purpose was to give Olcott a chance
to feel out the thrust of the new propeller with regard to
indicated airspeed, and to refamiliarize himself, after six
months, with the ground-handling qualities of the aerobody. It felt good. The new prop appeared to be delivering on its promise. Task 2 was, among other things, an

evaluation of the new vortex generators. Bill Putman, the aerodynamicist, observed with binoculars the hundreds of bits of yarn taped to the trailing edges as Olcott made three more runs without lift-off. "Those tufts are beautifully attached," Putman said. "Those vortex generators are working, no doubt about it. They have attached the flow." Olcott, in the cockpit, noticed that when the propeller got up to about four thousand revolutions per minute it developed a vibration and a peculiar buzz. Out of the north a Starlifter came and ponderously sank toward the runway. The 26 got out of the way, and the Starlifter was cleared for landing. Its ten tires screeched on touchdown. Almost at once, its big turbofans began to push for takeoff. The Starlifter went back into the sky. The Starlifter upset Miller. Jets that size leave roiling invisible whirlpools in the air for many minutes—turbulences that can destroy light aircraft. The morning air now smelled of kerosene and burned rubber. Never mind. If the little airship at the end of the runway was—in a general sense— to go up where it wanted to go, the Starlifters would scarcely last long enough to stink up another morning. The engine of the 26 was popping now and again, seeming to miss a little, and Olcott became concerned. All other considerations aside, he would have turned for the hangar and ended the day's outing. Olcott was beginning to respond, however, to pressures extraneous to the patterns of the testing. He could hardly be said to have adopted a devil-may-care attitude, but his basic approach—of cautious, methodical, incremental steps into the unknown— was giving way in fragments here and there. Olcott felt pressure, for example, in the rising skepticism of the personnel of NAFEC, and he felt a need to show NAFEC that the Aereon could get up off the ground higher than ever before. The popping sound disconcerted him, but he decided for the moment to ignore it. Task 4 was "lift-off out of ground effect"—a repetition of the final accomplishment of the flight tests of the previous September, when everything but Olcott had been thrown out of the aerobody in order to make it light enough to succeed. This time, though, the weight was back in the aircraft. The new propeller would pass or fail. Olcott asked for the tower's permission to take off and make a straight flight out of ground effect, landing at the far end of the runway. Cleared, he began to move—fire trucks, station wagons racing in parallel. The 26 took off at fifty-two knots and went up and out of ground effect with no strain at all. If there was turbulence from the Starlifter, the 26 punched through it without apparent shock. Olcott levelled off about sixty feet up, and accelerated to sixty knots. He had a margin of speed, power, and thrust that he had not had before. The engine popped again. He thought he felt a jolt, an interruption of thrust. Nonetheless, the flight continued and ended in steady trim. Olcott taxied in. Debriefing, in the cafeteria, he said, "I really wish we knew more about that engine. I really wonder if, even in the best of all worlds, it is producing ninety-two horsepower. The go, no-go factors are not completely clear." Task 5, the next in line, was what Aereon had been waiting for for twelve years—"up-and-away flight, to evaluate climb capability and speed range and to obtain additional assessments of the vehicle's handling qualities."

"Tell me about Task 5 again," Putman said.

Olcott said, "It's a circuit of the field."

"I thought we were going to do some shallow turns first," Putman said.

"I think there's enough extra airspeed so we're not going to stall out the vehicle or get an excessive rate of sink when we start to turn. I think the maximum speed is going to be about sixty-two knots."

"We don't know how it will turn—we've been thinking straight for so long."

"We may have the turning radius of a big airplane."

"At least try a few turns first that cover the width of the runway."

Olcott nodded. "What I want to do," he went on, "is to circle the field and then go over through the speed course and get some data. As I said before, though, I really wish we knew more about that engine."

One thing they knew about the engine, Putman reminded everyone, was that it had only a few more hours left. They had agreed not to use it more than twenty-five hours, and the deadline was now pressing in.

"We're really hanging on to the end of that string, aren't we?" Olcott said. Lifting one hand, he pinched the end of an imaginary bit of string. "I thought I could feel the aerobody respond to the fact that the engine was cutting out," he continued. "This would be disconcerting if I were circling the field. Check it out carefully, Link. And really check that prop, that vibration. If we lost the prop, we'd probably shake the engine off, too."

"No question about it," Linkenhoker said.

I looked around the table—Linkenhoker, Weber, Miller, Olcott, Putman—and noticed for the first time that everyone had sky-blue eyes. There was something new in their manner: a little more verve and less deliberation, aspects of a rush, under pressures to perform—NAFEC impatient, engine time running out, the money sponge all but dry. The group was talking about problems that were possibly quite serious—insufficiencies of the engine, enigmas of the prop—but at the same time the decision seemed inevitable. It seemed swept in. Task 5 was next. That was that. Tension had continued, in recent days, between Miller and Olcott. Olcott had said again that the word was around that Aereon did not really want these tests, because they would be too definitive. "They say the Aereon Corporation is afraid to learn that its concept is a bust, and I'm beginning to think they're right," Olcott had said.

"Well, they're wrong," said Miller.

"Well, I'll be there at NAFEC to see if they're wrong," Olcott said.

Miller said, "You're being quite unfair to Aereon."

Olcott said, "I'm concerned about my professional reputation."

This concern had removed from Olcott, in some measure, the advantage he had had in being unconcerned, merely consultative, detached. He was troubled by more than the vulnerability of his name. He was worried for Miller, for Linkenhoker, for the people who had been most deeply involved in the Aereon project over the years. "They've got not only their time and money tied up in this, they've got their whole ego tied up in it. Suppose it proves necessary to tell a person like Miller that the machine just won't hack it, that it just won't do it? How do you tell him? Yet I may have to tell him something like that.

That's what I'm paid to do. It may not work within the laws of nature. Sheer wishing will not make something go. Wearing the hair shirt a little tighter just won't do. I'm paid to give him a completely unemotional, objective assessment of that aircraft, and that is exactly the way it's going to be. There's a high probability that the outcome will be negative. I do sometimes wonder what people like Miller are going to do if this thing fails."

Miller, now, at the cafeteria table, was wearing a black tie covered with small orange tigers. On the field, earlier, he had been wearing an orange-and-black Princeton baseball cap. Among all the other things Miller was, he was an old grad, an old Tiger, a glint in the eye of Annual Giving. That was why the aerobody was painted orange, and why its stripes and markings were black. The over-all name for the development of the 26 was Project Tiger. Olcott, looking out the window and over the main runway, seemed to be contemplating something that was some distance above the horizon. Miller was taking the opportunity to remind everyone about the crucial importance of security at this time. "To any question say, 'We're not at liberty to talk about our tests.' Fend off people with cameras."

Olcott turned back to the group. "Fix the thrust meter," he said. "Fix the control-position transducer so it doesn't slip. See why the engine is popping. Maybe new plugs are needed. I'd like a smaller parachute, an eight-pound chest chute. Inspect everything thoroughly, especially the prop. Investigate but don't change the tab gearing. Put a tape recorder in the cockpit if possible. The next time we go, we're going to try to do Task 5."

MILLER HAD LONG SINCE discovered the hole in the fence between religion and superstition. All week long, as he worried and as he watched the weather, he looked for omens. When he learned that a major reunion of American airship men would be held at Lakehurst on June 26 and 27 next, he shivered with hopeful presentiment. The aircraft registration number of Aereon 26, boldly painted on its side, was N2627. The weather forecast for Saturday, March 6th, was more than promising. All the mechanical work Olcott had asked for had been done. So word went out on Friday, through the Aereon answering service, that the test group should meet at NAFEC at five-thirty the following morning. Reaching into a pocket, Miller took out his daily appointment book. It was called the Success Agenda Seven-Star Diary, and it included a fortune message for each day. Turning the page, Miller read the message for March 6,

1971. It said, "The mocker's arrow turns back like a boomerang."

Jack Olcott and his wife, Hope, happened to be giving a dinner party March 5th, as they had on the night before the first lift-off, six months before. Olcott mixed himself an aquatini—water with an olive in it—and after dinner he passed cordials around and said to his guests, "I hope you won't consider me rude, and I hope I won't break up the conversation, but I have a very early morning appointment and I have to retire for the night." He got up at four, and took fifteen minutes to dress, choosing a blue blazer, a blue-and-gold button-down striped shirt, his royal-blue tie with fleurs-de-lis, gray flannel trousers, and a pair of defeated, broken-down loafers with flapping soles. In the blazer's lapel was a small set of wings, emblematic of his membership in the secret society of Quiet Birdmen. At four-thirty, he parked his car at Morristown Airport, and shovelled an aging snowdrift from the apron of a T-hangar. He rolled out a Beechcraft Travelair, climbed in, and took off. His route south passed above McGuire Air Force Base. The McGuire approach controller said to him, "Are you going home late or getting up early?" Olcott gave the controller a straight answer. Crossing the Pine Barrens, he ate a box breakfast that his wife had packed—crullers and coffee, meat-loaf sandwiches. At five-thirty, he raised the galactic blue lights of NAFEC.

The big hangar was crowded. The 26 was nestled like an orange-dyed egg between the wing and tail fin of a Convair 880, a four-engine commercial jet roughly the size of a 707 or a DC-8. Around the 26 was a nonagon of gold nylon cord, strung among nine wooden stanchions. Linkenhoker, inside the barricade, was finishing up the preflight inspection. ("I had one major thing in my mind," he said later. "How might I feel if through some fault in the aircraft it cracked up and we lost a man? This was the foremost thought in my mind the whole time we were down there. I know one thing now: I'll never be placed in a position where I have to take complete responsibility for a man's life again. The design was good, but, nevertheless, the over-all putting together of the aircraft was mine, and that presented a hell of a feeling, I'll tell you. I thought, Here we are using an unaccepted structure and an uncertified engine, and we have low prior knowledge of the vehicle's flight characteristics. It presented a rather dismal picture in my mind. Fortunately, we were so damned busy—the buildup to the tests was so great—that I didn't have much time to think.") Everything seemed right with the aircraft and its engine—eleven hundred and one pounds minus Olcott, center of gravity fifty point three five per cent, examined and ready to go. Linkenhoker began to remove the gold cord. John Kukon, who had no official role to perform, had got up in the middle of the night and come to NAFEC anyway, unable to resist seeing this particular outing. Olcott, now in his test-pilot clothes, slot pockets bristling with stubby pencils, was telling Kukon stories about experimental airplanes he had known and interesting troubles they had had. There had been one in India, for example, "with a classical aileron wing-bending flutter problem" that always developed at just so many miles per hour. Olcott would accelerate the plane until he got a nice, pronounced flutter going. With a high-speed camera he would take pictures of the flutter.

He also told a story about a plane that had recently crashed in a bizarre way, yielding three survivors. If these were parables, they were to Olcott himself subliminal. His manner was, as always, calm and precise. He asked Kukon what he thought about the popping in the engine. Small power plants like that were not unlike model-aircraft engines, about which there was very little that Kukon did not know. Kukon told him not to worry. The 26 had a two-cycle engine, like a chain saw or an outboard, developing a great deal of horsepower for its weight. Two-cycle engines run on combined gas and oil, lubricating themselves as they go along, and just the right amount of air has to be mixed with this fuel to produce maximum horsepower. If the mixture is too lean, horsepower declines, and—more important—the engine can develop too much heat and destroy itself. If the mixture is too rich, horsepower declines also, but the engine functions well. One sign of a rich mixture is that the engine occasionally, harmlessly, pops. Olcott was used to flying four-cycle engines, and that, of course, was another story altogether. Popping in a four-cycle engine could be a symptom of catastrophic trouble. With a two-cycle engine, though, the best ratio for the fuel-air mixture was just a little way over on the safe side of the power peak—popping now and again, like corn on a stove.

Olcott thanked Kukon and said he felt relieved of that problem. The tall, telescopic doors moved apart. The Aereon was rolled toward the breaking day. Emerging from beneath the Convair 880, the 26 seemed small to the point of absurdity, with its little chain-saw-type engine mounted above the rear like a horsefly sitting on the head of a pin. Minuscule beside the giant airplane, the Aereon was hard to imagine at full scale, but if it ever grew to its ultimate conceptual dimensions it would not be able to insert into this big hangar a great deal more than its nose, for it would be the size of the Hindenburg and the Graf Zeppelin placed together in the shape of a T, with superstructure filled in to form an immense rigid delta. A couple of dozen Convair 880s could fit inside it. Linkenhoker, standing on an iron stool, primed the Aereon's engine, and tugged at a blade of the new propeller. The engine eventually coughed, ignited, and racketed against the walls of NAFEC. Olcott closed the hatch, radioed for permission to move, and routinely went up Taxiway Bravo toward the head of the runway.

For stopwatch timing and for photography, Miller, Putman, and others were delivered by station wagon to various points on the airfield. Fire and crash vehicles were operating on both sides of the runway this time—yellow lights, red lights flashing everywhere. The morning was pale blue, clear, and fine. The sun was above the horizon, its light streaming to the west. "This is a day the Lord has made," Miller said. "Let us rejoice and be glad in it. It's just ideal. No wind. Dry. Clear. This day is a gift." Olcott was facing west, at the head of the runway. Cleared, he accelerated, rotated, and lifted into the air. He climbed to forty feet and levelled off. He tried a coördinated bank and gentle turn to the left. It went well. He did the same to the right. The 26 responded as he had expected it would. It did not seem to have a tendency to roll excessively. The roll damping was light. He got a promising sense of the roll-control effectiveness of the vehicle. Reducing power over the seven-thousand-foot markers, he descended, landed, and taxied to the turn-around block at the runway's western end. The brief flight had been Olcott's warmup. He now felt that he had the vehicle all around him. NAFEC had asked him to take off to the east if he ever intended to leave the airspace of the runway, because the terrain at the eastern end of the NAFEC reservation had a dirt road winding through it and was particularly accessible to fire trucks and crash vehicles. Olcott was now facing east. He called the tower and said this was Aereon 2627 requesting permission to lift off and make a circuit of the field. Two miles of broad runway reached out before him. The parallel taxiway, where the fire trucks would race him, was to his right. To the right of the taxiway and a little more than halfway down was the great dark block of the NAFEC hangar, its near wall lined with tiers of offices behind shining plate-glass windows that reflected the low rays of the sun. The fire trucks and other cars were lined up and ready. Olcott showed them a raised thumb, moved the engine up to four thousand revolutions per minute, and left the head of the runway. He watched the black tape marks on his wind-shield, rotated, established his angle of attack, and went into the air. He climbed to forty feet. This time, however, he did not level off. Still in a position to abort the flight with ease, still in an environment he had been in many times before, he now had to decide whether the rate of climb was sufficient to warrant an advance to where he had not been. He had hoped for a rate of climb of two hundred, or even two hundred and fifty, feet per minute, but the engine was running flat out and he was getting a hundred and fifty. He figured that a hundred feet per minute, or less, would be so marginal that he would have to go down. This was, for sure, the inverse frontier—an exploration of the lower, most economical limits of aerodynamic possibility. Some commercial jets climb six thousand feet per minute. He watched the rate-of-climb indicator. It was holding at one hundred and fifty, positive rate of climb—positive enough for him to decide to stay with it. He put in a little rudder and made a slight right turn. Moving obliquely, he would add something to the time when he would be near enough to the runway to get to it if the engine failed. He was flying directly toward the NAFEC hangar, however, and NAFEC sternly told him to head somewhere else at once. He was about eighty feet in the air. If his engine failed, he could not have hit the NAFEC building even if he tried. The building was half a mile away. The 26 had a glide ratio of about five to one. From that altitude, the 26 could not have glided more than four hundred feet before scraping the ground. Moreover, the 26 was so light that if it had hit the building head on it might have had difficulty breaking the glass. Olcott corrected his turn, though, and continued to climb slowly to the east. It was like driving a station wagon stuffed with cordwood up the side of a mountain in first gear. He was getting there. He knew he would make it over the hill. Meanwhile, there was nothing to do but be patient. He reached a hundred feet, a hundred and fifty feet, two hundred feet, all the while reminding himself: Do not change anything. Stay at this airspeed. Hold the controls with constant pressure. Let the vehicle do the work. These tests are important, they must be concluded.

You knew all along that the vehicle was never going to behave like a homesick angel. It just wasn't going to climb like that. Within the margins of our considerations was a poor rate of climb.

The 26 was almost over the end of the runway, and was two hundred and fifty feet in the air. Seen from the ground and from a mile behind, it appeared to be a small black diamond moving into the sun. "Fantastic!" John Kukon said. "It's got a lot higher nose attitude than I expected, but if that's the way it is, so be it."

Olcott was now about to try the first significant turn the 26 had ever made. He could not with certainty predict what would happen. He did not have the altitude he had planned for, and he had to ask himself a lot of questions. He had to keep looking for and selecting places where he might set the Aereon down if the engine stopped, or if much of anything else went wrong. You can't wait until the engine quits to decide how to handle the situation. It's too late then. You have to know what you're going to do before you have to do it. So you are continually saying to yourself: What will I do if this happens? What will I do if that happens? If the engine quits now, I'll put the stick forward to make sure I'm going downhill, like the boy on a bicycle who doesn't want to pedal anymore. He's got to be pointed downhill or he'll topple over. Get the nose down. Establish the glide. Keep the airspeed the same, so you have control. Then go into one of those preselected landing spots. The engine is the primary consideration. Stability is the secondary consideration. The Aereon is not a broom balancing on the palm of your hand. It is a stable vehicle. Nevertheless, you do not yet know to what extent it is stable. As long as you don't disturb anything—as long as you move into any control input very slowly and smoothly—the chances are that you'll never upset the dynamics of the vehicle so drastically that you cannot cope with it. I am two hundred and fifty feet over the end of the runway. If the engine fails here, there is no way I can turn around and get back into the runway. If I were to try, I'd probably lose control of the aircraft. So what do I do? Where would I go? The dirt road. It is sort of a hard dirt road. I believe I could get in there. Maybe damage the nose gear but not do too much harm. I could negotiate that landing.

He went into the turn. He made it shallow, because he had never been in one before. His mind raced with the conditions and problems of the turn, addressing himself, addressing the aircraft. Let's take it nice and easy. Let's not depart too much from what we've done before. Here we go. This is the first time we've really got a sustained angle of bank. Really a turn. We know from the computer simulations that if the angle of bank gets a little too high, and the rudders are not coördinated just right, the vehicle will want to continue to the left and will be difficult to control. That would be disconcerting at low altitude. There's a straightforward way out, with use of rudders and manipulation of the stick.

He had taped one end of a bit of black yarn to the outside of the cockpit canopy, and now he watched it closely. Air should always be flowing straight back, no matter what maneuver the aircraft might be making. If the yarn were to move sidewise, the 26 would be going into a yaw.

The yarn was straight. The sideslip angle was zero—just what it was supposed to be.

The 26, continuing to climb, had turned through an arc of ninety degrees and was heading north. Olcott no longer needed the dirt road. If trouble developed now, he could probably get around to the runway, heading west, if he had to. It was like trying to cross a stream from one bare rock to the next bare rock, trying not to fall in. Meanwhile, in addition to and above all else, he was supposed to be collecting test data. What is the rate of climb now? What is the indicated airspeed? What is the angle of attack? What is the control-position transducer saying? How am I doing? How am I doing relative to what I want to be doing? How much will this turn hurt the rate of climb?

The 26 completed its wide arc to a hundred and eighty degrees and was headed west, parallel to, but considerably north of, the runway. The rate of climb had remained steady. The ship was four hundred feet up now, and it continued to rise until Olcott levelled off, as he had planned to, at five hundred feet. Data now flowed from the instruments. The maximum speed, full throttle, was sixty-four knots—a little better than Olcott had expected. He planned his route over the western end of the reservation, telling himself not to fly over the houses there, because that was not good professional technique in an aircraft that had a limited flight history and a configuration that had never flown before. I'll just have to go into the shrubbery if anything happens here, he told himself, but he swung into a perfect hundred-and-eighty-degree turn and was now pointed again into the sun. He was five hundred feet over the broad white stripes from which he had begun his takeoff. He had completed a circuit of the field.

"Wow!" Miller said, shooting straight up with his Nikon Super 8. "This is fantastic!" The fire trucks stopped running around. All the ground vehicles stopped. Everyone watched the sky.

Olcott now had the Atlantic Ocean spread before him, wide marshes and bays, the skyline of Atlantic City to his right, Absecon Bay straight ahead, and to the left the Brigantine National Wildlife Refuge. Almost below him was the Garden State Parkway, a superior alternative to the dirt road as an emergency landing strip. Northbound or southbound, the 26 could blend right in with the cars there, if necessary, at an identical speed. Olcott found the scope of his view extraordinary, because there were no wings around him to impede it.

Olcott again circled the field, this time reducing his airspeed to fifty-nine knots to see how the 26 would handle there. Then he went around again, at fifty-two knots, and again, at fifty. Each circuit was about eight miles.

"It's slowly sinking in," Miller said. "He's not going to come down."

In subsequent days, Olcott would fly the 26 right out to the end of its engine time. It would be tracked by NAFEC's theodolite, yielding, for two hundred dollars an hour, precise airspeed data. Olcott would do Dutch rolls and steady sideslips, kicking out hard with his rudders. He would do aileron rolls to the right, aileron rolls to the left, rudder kicks right, rudder kicks left, as if he were practicing swimming. He would climb, slow down, dive, speed up—a fundamental longitudinal mode, the phugoid motion. Investigating the phugoid, he would go into a steady sideslip and then "put in a doublet—just to get the thing excited."

"That's how we lost one of the Aereon 4s," Linkenhoker would say, biting a toothpick, watching from the ground, and then, perhaps because he was unable just to stay there and watch, Linkenhoker would jump into a Piper Cherokee and chase the 26 into the sky. I went with him. The 26 seemed to float beside us, over the field, pinewoods, the parkway—with tidal estuaries, salt marshes, and the sea beyond. Shafts of sunlight sprayed down from behind clouds in which the sun kept appearing as a silver disc, and, moving in and out of these palisades of light, the 26 went into smooth roll angles and controlled yaws—part airplane, part airship, floating, flying, settling in to landings light and slow. "Aereon is great," said NAFEC's chief executive officer. "Just look at it and you can see the potential. What made New York great? What made Chicago great? The carrying of freight." One could almost see New Yorks and Chicagos springing up under the slow-moving shadow of the Aereon as it flew. A subtler and perhaps more durable endorsement had come from NAFEC beforehand, however, on the day of the first circuit of the field. Flying on and on—the first circuiting flight lasted more than half an hour—Olcott looked up at one point to see a Starlifter approaching the field. The two aircraft—one weighing eleven hundred pounds, the other weighing seventy tons— were more or less on a collision course. "Tell them to give me plenty of room," Olcott said to the tower. "I cannot tol-

erate their wake." The tower told the Starlifter to turn
right, go south, and keep on going south indefinitely. "The
traffic on your left," the tower explained, "is an aerobody
—a wingless vehicle—proceeding northwest."

THE LAST TIME I saw John Fitzpatrick, I rode with him in his tow truck from Neshaminy Esso to a shopping mall, where he changed a flat tire for a woman who had telephoned him an hour and fifteen minutes before. When the call came, he had been adjusting a set of points in a two-year-old Mustang, and he had gone back to work on them and then had drifted off into a long elegy on the grace of naval airships. Finally, he shut the hood of the Mustang and remembered the woman at the mall. Even as his truck approached her, he was strafed with bitter complaints. Fitzpatrick said to her, "When something like this happens, the person it happens to tends to be conscious only of his own problem." Kneeling, he began vigorously to spin nuts. Driving away, he said to me, "That woman and I have no common basis for communication." Before long, Neshaminy Esso had a new owner. Fitzpatrick returned to New Jersey, to a job at Autobahn Motors, on Route 1, near Princeton.

The last time I saw Monroe Drew, he showed me four small sketches he had drawn of a rigid airship to be called Aereon 500. He said he had designed it to run on solar heat caught in a parabolic mirror. Its configuration consisted of four cylindroid hulls bunched together like four sticks of dynamite within an over-all airframe that was deltoid in shape. The lifting gas would be not helium but hot air, also produced by the parabolic mirror, except on cloudy days, when an alternate heat source would be required. Powdered-graphite spray coating would insulate the inside of the hull against the solar heat, which would be thermostatically controlled at about five hundred degrees. Among other things, Drew said, Aereon 500 could serve as a flying bakery. He suggested strongly that he might still bring force to bear that would unseat Miller as president of Aereon. "Bill pays lip service to Solomon Andrews and the airships, but Bill is just not an airship man," Drew said. Drew had become Senior Training Adviser for Medicaid in the State of New Jersey. The Fourth Presbyterian Church of Trenton no longer existed.

Jürgen Bock, the German physicist who developed the deltoid configuration with John Fitzpatrick and the Valley Forge computer, had been far more interested in theory than in experimental trivia, or so Miller once told me. Bock had lost interest, Miller said, and had gone home to Germany. Charlie Mills, the last time I saw him, removed from his desk drawer a letter from his friend Jürgen in Germany showing sketches of hypersonic lifting bodies that Bock was developing for a company called Erno, G.m.b.H. Charlie Mills, once Air Operations Officer for the naval airships at Lakehurst, was still teaching German at Hamilton Township East High School, outside Trenton, and was still using Windex to clean his glasses. He put them on, and looked over my shoulder at Bock's sketches. "If Miller ever saw these things, he would die," Mills said. Bock's new aircraft were wingless. From the side, they looked like fat pumpkin seeds. From above, they were deltoid.

John Kukon, in the Long Track at Princeton, was testing the control effectiveness of a model helicopter he had built with coaxial rotors. They counterrotated. There was no rotor at all on the tail. He was also trying to defeat problems associated with tilt-wing airplanes when they come close to the ground. Lift force decreases about twenty per cent and controls get soggy, and the eighty-five-thousand-dollar tilt-wing model that Kukon made was apparently not immune to these difficulties. What Kukon built for the university between nine and five was only part of what Kukon built. He was building all the time, and much of the time he was at home. There he had become interested in the problems involved in making model airplanes so light that they would almost be lighter than air— airplanes with fuselages two feet long and wingspans greater than that but with an over-all weight of one gram, or less than one-third the weight of a penny. When Kukon first made such an airplane and picked it up, he could not feel its weight on his hand. When he walked with it, he had to use a hesitation step, or the air the model was moving through would crush it. These aircraft were almost lyrical in their debt to structure. They were made of balsa wood twelve one-thousandths of an inch thick, had wing coverings of microfilm one ten-thousandth of an inch thick, and were held together by tungsten bracing wires too slender to be seen in most light. Yet they supported the torsion and tension of long rubber bands wound two or three thousand times. Propellers turned so slowly they could be strobed by a human eye. They turned one revolution per second or, toward the end of a flight, one revolution every two seconds. Cruising speed was just over half a mile an hour. Rate of climb was about fifteen feet per minute. International championships were held for indoor models of this type once every two years. The 1970 championships took place in a salt mine in Rumania, the 1972 championships in an airship hangar in England—big rooms with still air, where the object of the competition was simply to remain airborne. Kukon for many years had ignored this field, because he viewed it as more a matter of structure than of aerodynamics, since there was no limit on how light a model could be. It all seemed too simple. The lightest structure would fly the longest. Then, in 1971, the Fédération Aéronautique Internationale decided that indoor models in this category should weigh a minimum of one gram (the rubber motor was not included), and the rule change sent Kukon into his basement, for now it was a matter of structure *and* aerodynamics—something complex enough to be worth the while. He ground his own rubber cutters and cut his own rubber. He sanded down his balsa wood until it was translucent. He outlined the wing structure in strips of balsa that were thirty one-thousandths of an inch square in section. Onto still water he poured cellulose nitrate, which formed a film like an oil slick. Floating on the water, the film dried. Very carefully, Kukon lifted it with the aid of a wet wooden frame. A week later, after the film had stabilized, he placed it across the balsa structure of the wings—wings so frail they would have drooped like basset ears if they had not been held up by the tungsten wires. Propellers were similarly outlined in balsa and covered with film. The film was so thin— one-tenth the thickness of Saran Wrap—that light could not pass through it in the way that light ordinarily goes through a transparent substance. Instead, it refracted, reflected, caromed wildly, and split itself into all the colors of the spectrum in shimmering iridescence. When these airplanes flew, they were fantastically beautiful, slowly circling, climbing, spraying color in all directions. They flew, most notably, in Hangar No. 1 at Lakehurst—a giant barn a thousand feet long, almost two hundred feet high, steel-structured, sheathed in wood. This had been the hangar of the Hindenburg, which had burned just outside. Hangar No. 1 had been built for the big rigid airships, and now, in their continuing absence, the all-day twilight of the hangar was sometimes weirdly alive with eight, nine, or even ten almost invisible airplanes climbing slowly toward the roof, each barely heavier than air. In folding lawn chairs, the model-builders sat below. They held stopwatches. Around their necks were magnifying goggles, which they needed in order to see parts and wires as they prepared their models for flight. At launch, they carried their planes as if they were holding nitroglycerin. Sometimes an accident happened. There was a weakness in the fuselage, say, and the tight, knobby wound-up rubber band would snap the plane to bits. All the wreckage—of a model more than two feet long and two feet in wingspan—could be placed in the center of the palm of one hand. Flight occurred in three parts: the climb, the cruise, the dead-slow, hanging descent. The propellers turned like the arms of ballet swimmers, throwing off flashes of color. When the planes rose above a hundred and fifty feet, they were so hard to see that the competitors followed them with binoculars. If the aircraft got into trouble in the high girders, helium-filled balloons were sent up on long fishing lines and with the balloons the fliers expertly moved their planes into new courses of flight. The competitors brought their own helium to Lakehurst. For a time, Kukon, like everyone else, built single-wing airplanes. He suspected, however, that he could probably build a better airplane in a radically unconventional configuration. With the help of another engineer in Princeton, he worked out an optimum design in terms of wing area, wing loading, gross weight with rubber, ceiling height, a revised structure, ratio of motor weight to over-all weight, and forward speed. Many times over, he built his new configuration. His airplanes now had two wings, one forward, one aft, each with tip dihedrals. His airplanes were strong enough to carry more rubber than any competitor carried. He gave the rubber thirty-one hundred turns. The others always stopped at two thousand. The others included a building contractor, a five-and-ten-cent-store manager, assorted mechanical engineers, a welder, a retired Navy commander. They were empirical builders. Most of them had been flying indoor models for twenty and thirty years. When Kukon, a novice, arrived at Lakehurst with his awkward, outsize, two-wing aircraft, he inspired half-hidden chuckles all the way down a thousand feet of floor. He flew six, nine, fourteen minutes at first, less than anyone else there. He kept improving his techniques of building and flying, however, and eventually his models were flying twenty-nine minutes at a time. But that would not do. His competitors were five to seven minutes better than he. Giving up his unconventional aircraft, he reverted to a single-wing design, and soon he was up there with the best of the others, making flights of thirty-four and a half minutes, nudging into the zone of record time.

For twelve years, Everett Linkenhoker had literally welded together the fixed assets of Aereon—two prototype rigid airships, the one rampantly empirical, the other shaped by computer, the one blown over by the wind, the other successfully tested. As the test program ended, he knew that it signified an end of things for him, too. "I feel simultaneously the success of it all and the emptiness of the beginning of the end. In my mind are two conflicting trains of thought, a buildup and letdown all at one time. It unsettles me. I have to go out and rearrange my whole life." He was a welder, not an aircraft plant. He could hardly build a big rigid Aereon. He was an airship rigger. There were no airships to rig. He removed the 26 from NAFEC and supervised its storage—first, and briefly, at Red Lion, and then in a hangar at still another obscure airfield in south-central New Jersey. For weeks, he stayed close by the 26 and worried over what he was going to do. He stayed so long that he became maintenance manager at the airport. He completely hid the 26 from public view under an enormous black tarpaulin.

John Weber, the structural specialist, had an analogous problem, and it led him to a far more exotic solution. His license from the Academy of Aeronautics and his advanced degree in aeronautical engineering suddenly added up to nothing much but unemployment. Aircraft development had become virtually a no-opportunity field, forsaken by the federal government. Weber did not need a skywriter to spell this out for him. Bethpage, Marietta, Santa Monica, and Seattle had turned into burning ghats for his kind. So he bought into the Charles H. Dennis Cesspool Corporation, Fifth Avenue and Sunrise Highway, Bay Shore, New York. Readily, he developed expertise in the installation and cleaning of cesspools and septic tanks. Occasionally, he moonlighted for the Federal Aviation Administration, inspecting airplanes.

Bill Putman was working on advanced helicopter concepts and vertical-takeoff-and-lift projects at Princeton University. He had designed some of the helicopter models that Kukon was building there. Government contracts had atrophied so rapidly that the university's aerospace department was clawing hard for survival. After contemplating where the money might have gone, certain senior professors had developed serious interest in air pollution, or, as they called it, "aerothermochemical processes." A name change was pondered. The Department of Aerospace and Mechanical Sciences was thinking of becoming the Department of Aerospace and Ecology.

John Olcott had been assigned to explore "new-business development" for his firm, Aeronautical Research Associates of Princeton. He was seeking ways for the company to make its abilities negotiable beyond the confines of aeronautics. Meanwhile, he continued his contracted research and consulting work, test-flying anything interesting that came along. For the National Aeronautics and Space Administration, he was evaluating the use of spoilers on general-aviation aircraft. For Bell Laboratories, he had been flying a Cessna 337 with special wing-tip antennae for measurement work involving Safeguard antiballistic missiles. He had also been asked to evaluate a prototype two-engine low-horsepower airplane developed with private funds by a free-lance entrepreneur. Rapidly, Olcott's role in Aereon had thinned, the job for which he was hired having been completed in all but its consultant aftermath. When Miller had passed around finger-press-appliqué self-sticking tigers for everyone to wear after Aereon 26 had made its initial circuits of the field, Olcott smiled amiably, accepted one, and set it on a table. "This has been a very good day," Miller said. "This is a good moment." Olcott smiled, and said nothing. "I'm pleased with the way things turned out," he said later. "And I'm also pleased for the other people who are much more emotionally involved with Aereon than I am. To me, the challenge was 'Can you plan something with due consideration for all the risks, all the contingencies, all the limitations on the problem? Was your planning accurate enough? Were you good enough in your planning to be able to implement it without too many deviations?' There's no point in jumping up and down for joy. It *was* a pleasure to know that I had done my task professionally. The flight went as planned. We achieved all the objectives we were looking for. The implementation was about as complete as one could have hoped for."

Miller alone continued to work full time for Aereon. He still lived in a garden apartment off U.S. 1, and led Bible-study groups at Princeton University, and in the summertime worked as a volunteer for the Children's Sand and Surf Mission. The tests of the deltoid Aereon had been determinative, but no one could say in what sense. An experimental aircraft had at one moment been doing Dutch rolls and controlled phugoid oscillations in the here and now. In the next moment, it was a museum piece hidden away in a secret hangar. The group that had been around it dispersed. Months went by. A year went by. Two years. Aereon attracted interest but no developmental contract, no developmental funds. The summary result of all tests, all flights, all briefings and debriefings, all computations, two configurations, three propellers, one founder, four presidents, twelve years, nearly one and a half million expended dollars, and a hundred miles of circuit flight had been reduced to data that could be expressed on a single sheet of paper. Miller travelled around the country holding up the data like a lamp.